Native Food Plants of Texas

An Austin Forager's Guide Based on Indigenous Knowledge

by Cyrus Harp

Paleo Foraging
Austin, Texas

Copyright © 2025 by Cyrus Harp
Published by Paleo Foraging, Austin, Texas

All rights reserved. No part of this book may be reproduced, distributed, or transmitted in any form or by any means, including photocopying, recording, or other electronic or mechanical methods, without the prior written permission of the publisher, except in the case of brief quotations embodied in reviews and certain other noncommercial uses permitted by copyright law.

For permission requests, write to the publisher at:
Paleo Foraging
paleoforaging@gmail.com

ISBN: 979-8-9934846-0-0

DISCLAIMER

This book is intended for informational and educational purposes only. While every effort has been made to ensure accuracy, the author and publisher make no guarantees and assume no liability for any errors or omissions. Foraging, preparation, and use of wild plants carry inherent risks. Readers are solely responsible for identifying plants correctly, ensuring safe preparation, and complying with all local laws and regulations.

The author and publisher are not responsible for any injury, illness, or adverse effects that may result from the use or misuse of information contained in this book. Always consult multiple reliable sources and, where appropriate, a qualified professional before consuming or using any wild plant.

TABLE OF CONTENTS

INTRODUCTION	3
Preface	3
Author's Background	4
Historical Background and Sources	6
Indigenous Peoples	8
Warnings and Ethics	10
MAPS	12
Historic Natives of Texas	12
Regions of Texas	13
METHODS	15
Gathering	15
Processing	18
Cooking	23
Storage	26
Glossary of Terms	29
PALEOLITHIC PLANT FOODS	32
SPECIES ACCOUNTS	40
Overview	40
Fungi	41
Lycoperdon (puffballs)	41
Pleurotus (oyster mushrooms)	44
Trametes (turkey tail)	46
Conifers	47
Juniperus (juniper)	47
Cacti	52
Cylindropuntia (tasajillo)	52
Opuntia (prickly pear)	55
Trees and Shrubs	62
Acer (boxelder)	62
Sambucus (elderberry)	65
Rhus (sumac)	68
Ilex (yaupon)	73
Berberis (agarita)	77
Ehretia (anacua)	79
Celtis (sugarberry)	81
Diospyros (persimmon)	84
Arbutus (madrone)	89
Cercis (redbud)	91
Neltuma (mesquite)	93
Parkinsonia (paloverde)	108
Quercus (oak)	110
Carya (pecan)	126
Juglans (walnut)	131
Lindera (spicebush)	135
Morus (mulberry)	136
Forestiera (stretchberry)	141

  Fraxinus (ash) _____ 143
  Phytolacca (pokeweed) _____ 145
  Sarcomphalus (lotebush) _____ 147
  Prunus (plum and cherry) _____ 149
  Populus (cottonwood) _____ 158
  Sideroxylon (gum bumelia) _____ 160
Vines _____ **163**
  Rubus (blackberry) _____ 163
  Smilax (greenbrier) _____ 167
  Vitis (grape) _____ 173
Herbs _____ **181**
  Amaranthus (amaranth) _____ 181
  Chenopodium (goosefoot) _____ 187
  Asclepias (milkweed) _____ 192
  Liatris (blazing star) _____ 197
  Monarda (beebalm) _____ 200
  Oxalis (woodsorrel) _____ 203
  Plantago (plantain) _____ 207
  Capsicum (chiltepín) _____ 210
  Urtica (nettle) _____ 213
  Allium (onion) _____ 216
  Tradescantia (spiderwort) _____ 222
Wetland Plants _____ **224**
  Schoenoplectus (bulrush) _____ 224
  Nelumbo (lotus) _____ 228
  Typha (cattail) _____ 231
Asparagus Family _____ **236**
  Agave (agave) _____ 236
  Camassia (camas) _____ 246
  Dasylirion (sotol) _____ 248
  Nolina (beargrass) _____ 252
  Yucca (yucca) _____ 255
Palm Family _____ **261**
  Sabal (palmetto) _____ 261
Grass _____ **265**
  Poaceae (grass family) _____ 265
Toxic _____ **273**
  Cicuta (water hemlock) _____ 273
  Conium (poison hemlock) _____ 273
  Colocasia (taro) _____ 274
  Datura (thorn-apple) _____ 275
  Dermatophyllum (Texas mountain laurel) ____ 275
  Toxicoscordion (deathcamas) _____ 276
  Toxicodendron (poison ivy) _____ 277
INDEX OF SPECIES _____ **279**
REFERENCES _____ **284**
 Index of Peoples by Ethnographic References __ **284**
 Ethnographic and Historical References _____ **289**
 Botanical and Other References _____ **297**
 Digital Databases _____ **302**

INTRODUCTION

PREFACE

A thousand years ago in Central Texas, people met all of their needs with wild natural resources. Food, medicine, and materials were entirely derived from the surrounding environment. This fostered reverence for nature and a deep connection to the land and its species. What plants sustained them, and how were they used? The truth is largely lost to time. But I have spent the past decade trying to piece together a representation of the ancient lifeways of this land.

The history of Indigenous peoples in Texas was written with scraps of ethnobiological details buried in tomes of conquest and conversion. Those who persist, such as the Lipan Apache and Tonkawa, have struggled valiantly to maintain their cultural legacy. Yet for a representative depth of the lifeways of Indigenous foraging societies, we must look beyond just Texas, to the many Native peoples across North America who were documented while still living as hunter-gatherers, some of whom had not encountered Europeans until the mid-19th century. Over the years, I have combed through tens of thousands of pages of ethnographies, historical texts, and archaeological papers, searching for references to plants that also grow in Texas, in an effort to reconstruct the material lifeways of pre-contact Native peoples here.

The central question underlying this book is: what plants did peoples in the Texas Archaic eat, and how? To answer this, I not only studied the Indigenous records but also replicated Paleolithic and traditional methods of gathering, processing, and preparing wild plant foods. This book is the polished jewel of many years of research, experimentation, and refinement. Driven by a deep love for the nature of my homeland, and empowered by both scientific and Indigenous knowledge, I offer it as a map to help re-evaluate our relationship with the landscape.

I hope it inspires you to see this land differently, value it as those before us once did, and recognize what is at stake as its destruction accelerates with devastating effect.

AUTHOR'S BACKGROUND

In 2012, as an NSF doctoral graduate research fellow, I discovered a vast trove of ethnographies on Indigenous hunter-gatherers in North America in the library stacks at Berkeley, and it changed the direction of my life to create this work. I realized the best way to determine the practical uses of wild plants is to examine their historical uses by Indigenous peoples. At the time, I was studying ethnobotany and ethnoentomology, and began this book as part of that research. I was in California, the state with the best publications on historical Indigenous culture, as it was one of the last to be colonized. The university library was full of texts by anthropologists giving detailed accounts of the contemporaneous use of wild species by people still living in foraging societies.

I saw the opportunity to create evidence-based foraging when I realized that any useful plant will have been used by Indigenous peoples, and the more widely it was used, the more practical it might be for modern foragers. Reading a detailed account of the Miwok rendering massive stores of acorns into edible meal with only stone-age technology, I realized that I could apply the same method to oak species in Texas. And generally, I recognized lots of species as also occurring in Texas or being closely related to Texas species and realized they could be used in similar ways.

At the time, I had already read dozens of foraging books, but most catered to casual interest rather than practical subsistence. They often emphasized common introduced species or those with broad ranges, at the expense of more useful local natives. So, I began combing through every reference I could find on historical Indigenous uses of wild plants, taking detailed notes for every species. I included other practical uses, such as plants used for medicine and materials, subjects in which I also had interest.

I spent several years doing this before returning to Texas, where, as a native Austinite, I was more familiar with the local species, having been trained as a biologist there and foraging from an early age. I improved my research skills, refined my methods, and spent a decade taking notes on the uses of every plant genus that occurs in Texas. These references include many from Texas, but are from across North America.

During this time, I also practiced as many historical foraging skills as I could manage. My hands-on study of Paleolithic technologies has strongly informed my foraging and ethnobotanical knowledge.

To build an audience for this book and others based on my research, I began sharing my research, foraging experiences, and demonstrations of Paleolithic technology on YouTube and social media. I filmed replications of historical Indigenous foraging methods and edited them into informative videos, which can be found under the name "Paleo Foraging."

In this book, the "Notes" section of each species entry shares what I have personally learned that is most relevant to a forager. In "Character," I provide a brief description of the plant's salient characteristics for identification and use. In "Season," I detail when each plant resource is available. Seasonality varies by latitude, with stages of plant life cycles generally occurring earlier at lower latitudes. Readers further north can assume the season will be later, roughly proportional to the latitudinal distance. In "Nutrition," I present bromatological and phytochemical research on the plants, highlighting their food and health value. Finally, in "Practice," I describe my experience foraging them for food, usually by applying historical methods I learned in my research, along with efficient tips for modern foragers.

Unless otherwise noted, all photographs in the book were taken by me in Austin, Texas. All illustrations are also my own.

The author showing a day's foraging haul in early summer 2025 (anacua, mesquite, wild grape, wax mallow, paloverde, juniper, lemon beebalm, & more), mesquite drink in hand.

HISTORICAL BACKGROUND AND SOURCES

The research for this book draws from many reliable sources, especially ethnographic accounts of Indigenous North Americans from roughly 100 years ago, as well early expedition journals and other historical documents. The conclusions are limited by the dataset, as it cannot be assumed that ethnographers or historians were consistent or comprehensive in their records. Of the references cited, 41 cover Indigenous peoples of Texas, 44 are from the Southwest, 24 are from California, 15 are from the Southeast, 10 are from Mexico, and 36 are from other areas of the United States and Canada. In total, I considered more than 1,500 references, and read hundreds of thousands of pages in the process of research for this book.

One foundational principle guided my work: Indigenous peoples in one region likely used plant species in ways similar to those in other regions, particularly when a plant had a common or important use. By compiling records of Indigenous plant uses across North America, I was able to reconstruct a plausible Paleolithic ethnobotany of Texas. A further extension of this principle is that closely related species may have been used similarly. For this reason, my research includes not only direct accounts of Texas plants but also the documented uses of species within the same genera. If many species in a genus were consistently used in a particular way, it is reasonable to infer that other local species could have been used similarly. For example, the berries of many *Juniperus* species were eaten. Although no direct ethnographic evidence exists for the use of *Juniperus ashei* berries, their use can be reasonably inferred based on their similarity to other species, my own experience eating them over many years, and their continued use by modern foragers.

The central question guiding my research is: "What are the practical uses of Texas plants for a Paleolithic culture?" Ideally, this question would be answered by consulting the people who lived in Texas before 1528 AD, after which European-introduced diseases began spreading rapidly, often far ahead of direct contact, through vectors such as feral swine. These epidemics reduced Indigenous populations to a fraction of their former numbers. Texas was subsequently further ravaged as the frontier of the Spanish Empire and as a frequent destination for European ships along the Gulf Coast. By the late 1800s and early 1900s, when anthropologists began documenting Native cultures in earnest, the Indigenous peoples of Texas had been greatly reduced, displaced, and transformed.

Yet many Indigenous peoples elsewhere in North America continued practicing their ancient traditions well into this period, and many do so even today. Their knowledge forms the backbone of this book. In addition to ethnographic and historical accounts, I have drawn on archaeological studies, Indigenous-authored publications, and my own nearly 20 years of experience in biology. This background allowed me to identify each plant as accurately as possible, provide accepted scientific names, list synonyms and common names, and determine whether and where each species occurs in Texas.

This book focuses specifically on the food uses of plants in Austin and Travis County. Limiting the scope in this way makes the book both more practical for the general public and more personally grounded, since I have direct familiarity with the species here. Austin lies at the intersection of the Southeast, Southwest, and Great Plains. Its subtropical climate allows Travis County to support more plant species than all of New England combined. Most of these species also occur elsewhere in Texas, and many are widespread across North America. Thirty-three accounts in this book summarize genus-level uses, meaning that readers from the Eastern United States, the Great Plains, the Southwest, or northern Mexico will also find much relevant information.

Ilex vomitoria
yaupon
- *boil leaves for caffeinated tea*

INDIGENOUS PEOPLES

This work makes a deliberate effort to credit the Indigenous knowledge systems that form the foundation of this subject and treat the associated cultures with respect. With few exceptions, all knowledge of wild edible plants, the basis of every foraging book, originates from Indigenous sources. Yet few foraging books acknowledge this fact, and when they do, the credit is often generic, referring only to "Native Americans" in general.

In my research and in this book, I name each of the Indigenous peoples individually whenever possible. This serves two purposes. First, it is to simply credit the people by name, as they deserve. Second, listing all the peoples by name that ate a particular plant or used a particular processing method clearly demonstrates the practical utility of that food or method. For example, one can see the obvious utility of mesquite beans for food by the long list of Natives that ate them. This book almost exclusively covers plants that were recorded as being historically eaten by Indigenous peoples. The paucity of sources describing the historical consumption of a plant or total lack thereof does not necessarily mean this plant is not a viable food resource, as it may be simply a reflection of the source material or the limited geographic range of a plant. But if a plant in Texas was historically an important food resource, it is likely to have been documented to some extent by the references in this book.

The Indigenous peoples mentioned have been identified as accurately as possible with modern English names, or at least by the area in which they were living. Note that exonyms, names given by outsiders, are generally used. The endonyms, the names peoples use for themselves, are used in some cases where the exonym has been reportedly widely rejected or when the endonym is well known (e.g. Diné instead of Navajo).

Those named (with some rare exceptions) are still living peoples, but the uses are referred to in the past tense. This reflects the historical context of the sources and acknowledges that the practices described may not represent the current traditions or perspectives of the same peoples today. Although some modern references are included, this work is primarily a historical reconstruction.

Indigenous Plant Names:
I have left the special characters and diacritical marks as written in the original sources, or when ambiguous, defaulted to the closest character or mark in the International Phonetic Association chart.

Translations are also given as in the original source texts, following this format: Indigenous people: *plant name* – "translation" / *alternate plant name*, or *plant name* ~ "rough translation." When the translation is given in the form "__, __", the comma between denotes that is the direct order of words in the Indigenous name. For example, Seri: *coquée quizil* – "chiles, little (plural)" refers to chiltepín, and could be translated as "little chiles" but this form allows one to know the Seri word for "chiles" is *coquée*. When multiple names are given for a species, sometimes reflecting regional or vernacular variation, much like English common names, I have included them all.

Names may differ between bands of a people, have shifted over time, or not reflect the current name used. Often they were the names given by individual informants in each ethnobotanical account, transcribed by authors who may not have been fluent in the language. In some cases, a plant name may actually refer to a specific plant part used or the product made from it, rather than the entire plant.

Indigenous plant naming systems (folk taxonomy) can be more discriminating than scientific taxonomy. Varieties may have distinct names in Native languages though botanists consider them one species. Folk taxonomy, especially at higher levels of grouping, does not necessarily follow genetic relatedness or patterns of descent. They tend to be more understandable and relatable than the relatively abstract and somewhat arbitrary groupings of scientific taxa. For example, one folk taxon of the Gila River Pima, "*Shuudagĭ che-ed̦ Hai'ichu Vuushdag*," means "plants growing in or on the water."

Indigenous names are included for three main reasons. First, plant names in any language are relatively rare words, and many of these languages are at risk of losing them. Recording and preserving them is therefore crucial. Second, Indigenous names are often highly descriptive, offering insights into distinctive properties or uses of plants. Third, including these names credits the cultures and the Indigenous knowledge systems that produced understanding of plant uses, while also allowing readers to cross-reference ethnographic sources more easily.

Acknowledgments:

This book would not be possible without the knowledge generously shared by Indigenous informants, to whom I extend my deepest gratitude for their contributions to humanity. I also thank the ethnographers and historians who documented the historical lifeways of Indigenous peoples across North America, preserving knowledge that continues to guide and inspire.

WARNINGS AND ETHICS

The plant uses described in this book are presented for historical and educational purposes only and are not intended as recommendations for practice. Attempting to replicate them may pose risks of physical harm or legal consequences. Modern foragers should always follow ethical guidelines and remain aware of potential dangers.

By understanding and appreciating the historical uses of wild plants, readers can also find motivation for wilderness conservation. Foraging is currently illegal on almost all public lands. Many National Forests allow foraging to some extent, but each has its own regulations. I seek to show modern people the myriad practical uses of wild plants, animals, and minerals in order to give value to undeveloped land. Hunting and recreational uses of land, such as bird watching, camping, or ecotourism, are the main drivers of land conservation in North America and globally (Di Minin et al. 2016, Loveridge et al. 2009, Paulson 2012, Serfass et al. 2018, Stronza et al. 2019). Foraging can give value to the more commonplace public wildlands, thereby motivating their conservation. Nature should be valued for more than stunning scenery, camping, or hunting.

General foraging ethics guidelines I follow:
1) Do not harvest more than you need for yourself and family to eat.
2) Do not waste if it is avoidable.
3) Always leave plenty of fruits, seeds, and other resources for wildlife.
4) Do not take more than about 10% of what is available from a particular plant or area.
5) Do not harvest any parts that will kill the plant if the species is not common in the area. This requires you to be aware of the population health of the species you are gathering.
6) Do not harvest from a plant or area too frequently. Allow time for regeneration and for others to forage, applying these guidelines every season and year.
7) If you can assist in the dispersal or health of a native plant, do so when possible.
8) Always give thanks and leave an offering. It fosters an attitude of appreciation instead of exploitation, leading to conscious and unconscious valuation of the plant.

General warnings:

If you are not 100% sure of a species' identification, do not harvest it. It may be a deadly lookalike or a rare relative. I do not recommend practicing anything in this text.

There are other considerations I keep in mind when foraging. Be aware of what pollutants the plant has been exposed to. Pollution is rampant in our world. Even in remote areas, pesticides may have been used, or there may be soil contamination. For example, cities are known to inject glyphosate (Roundup) into the stumps of trees cut down for invasive species control, which can pollute surrounding areas. Some National Forests are sprayed with chemicals by plane. And even your back yard may have lots of old trash buried that leach chemicals absorbed by the plants. Roadside plants are exposed to exhaust and dust that may be harmful. Trailside plants may have excretions of humans or pets. Wildlife can harbor diseases that transfer to humans via excretions left upon plants, especially if the plant is not washed. That being said, the substances to which many plants that are found on grocery store shelves are exposed are often appalling.

An array of local foraged foods, some fresh and some dried, and a foraging net used to gather many. Cherokee corn is included. *Prunus*, *Diospyros*, *Carya*, *Vitis*, *Sambucus*, *Forestiera*, *Neltuma*, *Ilex*, *Berberis*, *Celtis*, *Juglans*, *Parkinsonia*, *Elymus*, *Juniperus*, *Rhus*, and *Opuntia* are all shown.

MAPS

HISTORIC NATIVES OF TEXAS

At least 124 tribes inhabited Texas from first contact through the Spanish Mission period. The highest density occurs along the Gulf Coast, in East Texas, and along the *Camino Real*, likely reflecting the greater number of expeditions that visited these areas. Each tribe's name is placed near the center of its estimated home territory. The image uses the modern state border as its frame, with major rivers highlighted in bold blue lines over an early satellite image to aid geographic orientation.

I compiled the names from various references (Anderson 1999, Carter 1995, Foster 1995, Foster 1998, Foster 2008, Newcomb 1993, Perttula 2012, Sturtevant 1979), to create this map, which is perhaps the most comprehensive of its kind for Texas.

REGIONS OF TEXAS

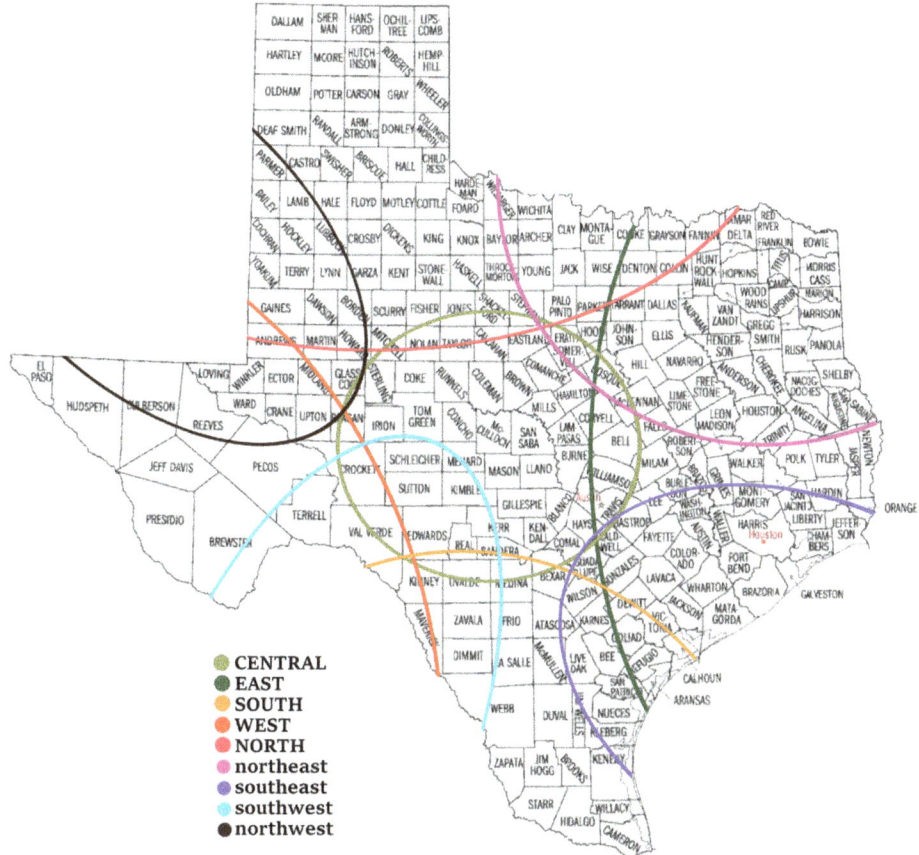

Regions of Texas, used to describe the distribution ranges in the text of species and genus accounts. Base Texas county map: Wikimedia.

These nine areas are defined for the purpose of describing plant ranges in the text of each species account in this book. The areas somewhat correlate to the Gould Ecoregions of Texas. There are five main regions, roughly corresponding to the four cardinal directions plus Central Texas, along with four additional regions approximately aligned with the intercardinal directions. All regions are defined to roughly correspond to floristic zones based on ecoregions, geography, and my analysis of thousands of plant range maps. All Texas counties are shown.

Location key:

Abbreviation	Meaning
C & E TX	Occurs in central and eastern Texas, as defined by above map.
C & far E TX	Occurs in central Texas and a narrow band of the easternmost counties.
E, C, & S TX	Occurs in eastern, central, and southern Texas (approximate relative abundance in that order).
all TX except S	Present throughout Texas except the southern region.
C & sparse SE TX	Occurs in central Texas and sparsely in the southeast.
all TX	Present across all regions of Texas (not necessarily in every county).

These abbreviations are seen following "Loc.:" (location) under the species and genus headings in the species accounts section.

The occurrence and rarity designations are based on data from the Biota of North America Program (BONAP), USDA Plants Database, World Flora Online (WFO) database, Kew Plants of the World (POWO) database, Global Biodiversity Information Facility (GBIF) database, Travis Co. herbarium specimens, observations on iNaturalist, and my personal experience as a lifelong resident of Travis Co. I have over 15 years of experience in professional and vocational identification of species.

METHODS

GATHERING

Overview:

There are a few gathering techniques useful to foragers historically and today. These include hand gathering, "strike and catch," and digging. Several tools are useful for gathering, and they can be easily made or obtained.

Hand gathering:

Hand gathering is best for larger plant products and is the default foraging method, being generally possible to use on most plants. This can refer to gathering directly from the plant or the ground beneath it. I often use it for wild plums, pecans, mesquite beans, and many other foods.

Strike and catch:

The "strike and catch" method is the most efficient way to gather small fruits from shrubs, or grains and seeds from grasses and herbs, and can be applied to many other foraging targets. In this method, one uses a tool to strike the vegetation, causing the plant to drop the fruits or seeds, which are caught in a receptacle vessel.

This striking tool can be a basketry racket, as was commonly used by Natives in California for harvesting seeds and grains, a similar implement, or simply a stick. I use a stout stick about an inch in diameter and three feet long. The receiving vessel needs to have a wide opening, and can be a large basket, a plastic bin, or a net-like implement.

I use a fabric net with a handle. This is an implement of my own design and manufacture. It is composed of a stiff hoop, to which is sewn a circle of duck cloth (heavy-duty cotton fabric). For the hoop, I bend a long shoot of green willow and wrap the ends together with cordage or use a piece of PEX tubing (common plumbing material). A six-foot section of ½ inch PEX can be bent into a hoop and both ends attached with a six-inch section of ¾ inch PEX.

The circle of cloth needs to be larger in diameter than the hoop, ideally about a foot or slightly longer, edge-to-edge, or about six or more inches from the edge of the hoop to the edge of the cloth all around the circle (shown below). This gives it enough slack to be sewn on to the hoop and also sag down, to form a shallow, net-like collection surface. I hand-sew the edge of the cloth to the hoop with a simple stitch. To this, I attach a stick (also about one inch in diameter) across the hoop, projecting a foot or two outward on one side. I cut small holes in the cloth to accommodate lashings I use to tie on the stick. It is important to attach the handle to two points on the hoop directly across from each other to provide stability.

Above: foraging net under construction. A bent willow shoot lashed into a hoop is on top of the circular cloth, cut the shown size larger than the hoop's diameter. Overlay: shows the edge of the cloth being stitched to the hoop, leaving slack to form a hanging collection surface.

Right: finished foraging nets. The top one has a willow hoop, povertyweed handle, and juniper stick. Overlay: one made with a PEX hoop and Texas persimmon handle and stick. The handles are lashed with thick thread or cordage.

Digging:

Numerous foraging targets, such as wild onion bulbs or cattail rhizomes, are underground. The universal historical implement for gathering underground foods was a digging stick. A long, pointed stick was the common tool to dig up edible roots, excavate root strands for basketry, and more. The points were sometimes fire-hardened. Transverse handles of horn, antler, wood, or simply the fork of the same branch were used. Hardwoods such as oak were preferred, and elk antlers were also sometimes used. They were sometimes flattened at the end instead of pointed. Similar tools were used to cut agave plants from the ground, with a sharpened, shovel-like, flat end.

The modern forager can easily adapt this method by cutting a stout, hard stick at a sharp angle to serve as an improvised trowel or shovel in the field. I put points on the ends of the sticks I use with my foraging nets for the "strike and catch" method, so that I can use also use them for digging. But a dead branch broken from a nearby tree is usually sufficient.

A hand trowel is also easily carried in a pack when seeking to forage underground foods. A sharpened trowel (e.g. a Japanese hori-hori tool) is a convenient and useful foraging tool for digging, cutting, and other applications.

Using a digging stick to excavate wild onion bulbs. Note that it is best to dig around the roots first, loosening up the surrounding soil, to prevent damaging the bulbs.

PROCESSING

Overview:

There are processing techniques useful to foragers historically and today. These include threshing, winnowing, sifting, flotation, water cleaning, drying, grinding, pounding, and pulverizing.

Threshing:

Threshing is a technique used to remove seeds from seed heads (infructescences) or whole plants. This was commonly used for grass grains and amaranth seed heads. The grasses or seed heads were cut and bundled and transported to a more convenient location, where they were dried. They were usually cut just before they would naturally drop the seeds, which usually occurs in the process of drying out as they mature.

Once dry and ready to release the seeds or grains, the bundles would be spread out and struck with a stick, paddle, flail, or similar device to agitate the seed heads and dislodge the seeds. This was done upon a surface such as leather, where the grains or seeds would fall to be collected. The larger plant pieces could be removed by hand and the rest of the debris was removed by winnowing.

Winnowing:

Winnowing is the process to remove debris from a plant product, usually to isolate seeds, grains, or fruits. The term is derived from the fact that wind was usually used to carry away the debris. A gentle breeze is often all that is needed, but the technique must be adjusted depending on the strength of the wind. I often use a box fan and adjust the setting and position depending on need.

The principle applied is that under a constant wind, lighter particles dropped from the same height are displaced farther than denser ones. The products (fruits or seeds), along with the associated debris, are placed on a basket or other surface. The material is scooped up in the hands, elevated a few feet above the surface, and allowed to gently cascade down. As it falls through the air, the wind pushes the lighter debris further than it pushes the seeds or fruits. Aimed correctly, the debris falls outside of the collection surface, while the denser product falls back onto it.

Historically, the material was held in a basket, tossed up, and the product caught as the debris was carried away. This requires more skill than the method of scooping with one's hands, but is more efficient. In general, it takes a bit of skill and practice, but this technique is extremely useful in processing foraged wild foods and is applicable in many scenarios.

Sifting:

Sifting is the process to separate or gradate particles of varying size. This is very simple when using modern tools such as flour sifters. These are containers with a wire mesh bottom, allowing only certain particle sizes and smaller to pass through. These come in different grades. I use these to process both mesquite and acorn meal. Baskets with open weaves of various apertures can be used similarly.

Sifting can also be achieved with flat or gently curved basketry trays. Historically, Indigenous baskets were specifically made for countless purposes such as sifting acorn meal in this manner. Basket sifting takes plenty of skill and practice to finely achieve, but a beginner should be able to at least crudely approximate it. The product, such as acorn meal, is placed on the tray, which is shaken, gently tossed, tapped, or agitated in a manner to cause the larger particles to move to one side of the tray or fall off, while the finer particles remain closer to another side.

A final method of sifting used historically by Natives in California for obtaining the finest grade acorn meal is to place ungraded meal into a basket, then dumping out the meal. Into the fine weave of the basket will have settled the finest grains, where they remain, even when the basket is upended. These grains can be removed by turning the basket upside-down onto a collection surface, then striking the bottom of the basket with a stick or bone tool, causing the fine grains to be released.

Flotation:

Flotation may be useful for separating acorns and nuts. Some modern foragers advocate separating acorns by flotation. The acorns which float are discarded. These are indeed usually acorns that have been infested with beetle larvae, were aborted, or are otherwise not viable for germination. It similarly works with pecans and walnuts.

Despite its utility, there are problems with this method. I consider it unnecessary, as acorns or nuts that are "duds" (aborted development) or are heavily infested are usually visually identifiable. I also usually process acorns with weevil larvae in them and eat them anyway. Acorns must be thoroughly dried before grinding them into a meal for leaching, so if one plans to do this, the flotation method adds drying time. Moisture is the primary cause of spoilage of acorns and nuts, so I rarely practice flotation.

Water cleaning:

"Water cleaning" is my own term for a particular process I use to clean fruits and remove associated debris. I place freshly-gathered ripe fruits into a large vessel of water to clean them, remove debris, and remove bad fruits. The debris and bad fruits tend to float to the surface, from where they can be skimmed off.

After this initial skimming, plenty of debris will remain suspended in the water. I remove this by applying a simple principle: debris sticks to my hands but the fruits do not. I repeatedly dip my hands into the water, allowing the debris to adhere, then sling off this debris. It takes about five to ten minutes to clear most of the debris from a large batch of fruits.

Some smaller debris will remain, but this can be removed by straining the fruits through a colander or coarse-weave basket, leaving nothing but the good, clean fruits. I usually then let the surface water dry off the fruits before any further processing.

Drying:

Drying is one of the most important historical methods for preserving foods, especially foraged wild fruits. Sun-drying was historically the most common form of dehydration. Despite the name, this does not require direct sun exposure. It is unclear in most historical accounts what "sun-drying" means exactly, but direct sunlight exposure is usually deleterious to food quality, especially that of fruits and vegetables, definitely lowering antioxidant richness.

My preferred method of sun-drying is to place the food between two flat basketry trays, weighing the top basket down with a rock. One can also build a sun-drying rack, which is more of an enclosed cabinet with mesh sides and a solid top. Drying in direct sunlight works, but the UV rays will bleach out or discolor the food. Enclosing the food also prevents insects from getting to it.

I built a simple drying rack with cedar posts driven into the ground and poles laid across the forks, with a surface of giant cane stems, all lashed together with grape vines, which has stood for years despite its rough construction. This elevated surface is better for drying as it improves air circulation and increases sunlight exposure.

Of the modern appliances most useful to a forager, a dehydrator is near the top of the list. A basic electric dehydrator can be purchased for around $50 and yields a superior product compared to sun-drying. It also precludes concerns about the weather. A regular oven can also be used for drying. Most ovens can be set to 200° F or below, which is adequately low for drying. However, one must be careful not to over-dry or burn the food when drying in an oven, and the product will be inferior to that from a dehydrator, which is typically set at 100° to 150° F.

Grinding, pounding, & pulverizing:

Historically, a mano and metate or mortar and pestle was used to reduce seeds, nuts, bean pods, dried roots, and other foods into meal. "Meal" just refers to the ground-up edible product. "Flour" specifically refers to the meal of grains, though many authors use these terms interchangeably. "Piñole" is meal, or a mixture of meals, eaten as is. "Cakes" are solid masses of meal that are often cooked.

A mortar and pestle was universally used to reduce a vast amount and variety of plant material into workable consistency for food, medicine, and material. Mortars ranged in size from being made from large tree trunks to being handheld, and pestles ranged in size from six feet long and a leg's circumference to a palm's length and thumb's circumference. Their dimensions were designed to be ergonomic and adequately fit the purpose. Some methods used multiple people alternately pounding with their pestles in one mortar. Mortars and pestles were most commonly made of wood. The mortar bowl was made by burning it out with embers and scraping out the char to shape it. Wood burls or knots were preferred for making mortars.

Mortars were also commonly made directly into the bedrock, including in limestone strata of Texas. Stone mortars were made by pecking and grinding. In pecking, one repeatedly strikes a softer stone with a harder stone to break out chips or grains. A chert or quartzite cobble is adequate for pecking out limestone mortars. A half day of work is required to peck out an adequately sized mortar bowl in limestone with chert. To put a smooth finish on the bowl, I use river sand, water, and a sandstone cobble. The sand serves as an abrasive, and a rounded cobble can be used to grind it into the bowl after pecking out its rough form. As the limestone grinds out into a slurry, water can be added to wash it out with a grass brush or wad of grass, exposing the fresh surface to be further ground. Water should be generally applied throughout this sand-grinding process to carry away the sand and limestone particles.

The mano and metate were both stone. The metate is a flattened, slightly curved surface. The mano (held in the hand) is equivalent to the pestle, but is often also flattened somewhat.

A range of sizes of ceramic or stone mortar and pestle can be readily purchased by the modern forager. However, finding very large sizes may be difficult. For this, one can use a carving gouge and maul to carve out the center of a log or burn it out.

Most modern foragers will likely find electric appliances to be preferable, as they are far more efficient. A sturdy blender is an indispensable device in my use of wild foods, and is my top choice for a single modern tool to aid foraging. A food processor or even a coffee grinder can also be used for many purposes of reducing foods into meal. There are also grain mills, which are better suited to grinding harder materials, such as paloverde beans. These are more expensive and I have not personally found them to be necessary, though one interested in certain foods may find them to be useful.

Above: Limestone mortar and chert cobble being used to grind acorn meal. Overlay: shows the process of pecking out the bowl with a hard chert cobble.

Right: Wood mortar and pestle being used to pound mesquite pods. The mortar was carved from a section of a black willow tree trunk. The pestle was made from a large mesquite branch that was de-barked, smoothed, and a slightly rounded bottom carved into it.

COOKING

Overview:

There are some cooking techniques and terms useful to foragers historically and today. These include boiling, decoctions and infusions, earth ovens, fire-roasting, and others.

Boiling:

Boiling is an essential process for food and medicine processing. Many different methods have been historically used by Indigenous peoples in North America.

Pottery vessels were placed sitting in the fire or suspended above it. Pottery can be made by pounding dry clay deposits into a powder, mixing it in a 4:1 ratio with sand, adding water until the mixture can be rolled into coils that can be bent around the finger without breaking, and forming this into a pot. The hands are all that are necessary to make coiled pottery. Form a circular base, roll out coils, then build up the sides gradually, pinching, smearing, and smoothing the coils into one another. Water is used to help smooth it and a wood or shell scraper can also be used. After the pot is completely air-dried, heat it by a fire to drive off remaining moisture. Then place it on quartz cobbles as a base and stack mesquite firewood all around it without touching the pot. Let the fire burn out, then remove the pot from the fire.

Stones, usually about the size of a fist, were heated in a fire and placed in a container filled with water, then food was added. The container was a rawhide basin, a waterproof basket, a pottery vessel, a wood bowl, or simply a hole in the ground. Note that many stones can explode when heated in the fire, especially those with pockets of water inside, such as a porous stone taken from a creek. Close-grained, porphyritic, igneous rocks (e.g. granite and basalt) and soapstone are historical examples. I have successfully used quartz and quartzite cobbles gathered from a hilltop to boil liquids this way. Two sticks, one with the end bent in a loop, can be used to remove the stones from the fire and place them in the vessel.

Bark from elm trees (particularly *Ulmus rubra*) was used to make a kettle by tying large pieces together at their ends with strips of inner bark. This kettle was filled with meat and water, suspended between two sticks over a fire, but not in contact with flames, and the meat was cooked before the bark was burnt through.

A decoction is the liquid made from boiling plant material in water, then straining it out. An infusion is the liquid made from pouring boiling water on plant material and straining it out.

Earth oven:

Earth ovens were a form of cooking historically used by Indigenous peoples globally. They used the principle of heat transfer. Stones were heated by a fire, food was placed on these hot stones, and both were buried in the ground to insulate them, which allowed the heat to transfer from the stones to the food. There were many different forms of earth ovens. In the simplest form, a pit was dug and lined with stones and a fire was built in this pit. After the fire burned down, the stones were swept and the food placed atop them and buried. The food was usually protected by layers of vegetation to prevent the ash and earth from contacting it. Sometimes, an additional layer of heated stones was placed on top of the food before burial. A fire could also be built atop the buried pit. In some cases, the fire may have been built first, then the stones placed on top of the fire, settling as the fire burned down. Earth ovens were often quite large to cook bulk amounts of food for many people or for storage.

Earth ovens used by the Apache for agave hearts were 10-12 feet in diameter and 3-4 feet deep. They were lined with large, flat rocks, with oak and juniper wood being used for fuel. The plant lining used was moist grass of various kinds, or beargrass, since it does not burn easily.

In a special form of an earth oven, the Haudenosaunee dug pits in the sides of banks or in a clay deposit, built a fire in them, removed the coals, placed food inside, and covered it with ashes.

I have successfully used earth ovens to cook large or bulky foods. I used quartz and quartzite cobbles gathered from a dry location. I burned juniper and mesquite wood to heat the stones, brushing off the coals and ashes to the side after it cooked down. I have used boxelder foliage, annual ragweed foliage, and damp grass as the lining materials. After burying this, I build a fire on top when I want to apply extra heat. I open the pit after about 8 to 12 hours, after the ground is cool to the touch. I am careful to not spill dirt into the food as I uncover the vegetation lining. This technique works quite well and, despite the long cooking time, does not allow the food to dry out.

The modern forager can mimic an earth oven by wrapping the food item in aluminum foil (to prevent moisture escape) and cooking in a conventional oven at low temperature (e.g. 200° F) overnight.

Other methods:
These include baking on a heated flat stone, searing directly on red coals, and broiling on sticks stuck into the ground. Placing food on a rock right next to a fire can also be used for cooking.

Baking in the ashes of a burned-down fire, with coals still burning underneath and plenty of residual heat, was a common method of cooking food without the possibility of burning it, similar to the earth oven.

Seeds were often roasted by shaking or stirring them in a basket along with live coals.

Left, top to bottom: Pinch-pot under construction; firing it; chunk of raw clay used and fired pot; boiling yaupon leaves in such a pot.
Right: Forked sugarberry stick bent into a loop used to place hot stones in acorn meal for boiling.

STORAGE

Overview:
    Historically, foods were stored in a variety of ways. These included platforms, granaries, caches, pottery containers, and hanging methods. There are some other storage methods suited to a modern forager.

Platforms:
    Elevated platforms were historically used by Indigenous peoples to store foods such as nuts, acorns, and seeds. Such platforms could consist of a simple wood frame, ten to twelve feet high, with a surface of canes secured by bark lashings on top, or even just a boulder. These platforms were often outdoor structures, but were sometimes used inside dwellings.

Granaries:
    Basketry granaries on elevated platforms were commonly used. A semi-spherical willow basket about three feet in diameter is one example, being used by the Cahuilla for mesquite pod storage. Insect infestation was prevented by methods such as sealing the granary with mud, using insect-deterring aromatic herbs (e.g. *Artemisia* sp.) to construct the granary, or mixing in fine sifted ashes. Granaries were built in many different ways, sometimes covered with bark that served as a roof. They were usually constructed in such a way to keep the stored food off the ground, out of the rain, and sealed from infestation.

Caches:
    Underground caches were often used, especially when nomadic peoples left an area and planned to return at a later date. For buried caches, the Wichita made a large hole that was wider at the bottom than at the opening, lined it with a bison hide, filled it with food, and closed the hide like a sack before burying it with earth and covering it with grass. Food cache holes were dug under cliff overhangs by the Ute. Sacks of rawhide or woven bark were lined with clean dry grass, the sacks covered with bark, grass, and rocks, then covered with dirt and more rocks, and a fire built on top to disguise it.

Pottery:
Pottery vessels were often used by some Southwest Natives to store food. Often, such vessels were large with a narrow opening, and had lids that were hermetically sealed. Sealants used included creosote lac, pine pitch, or other resinous plant saps, which were heated to liquefy and then spread over the vessel.

Hanging & other:
Pots, baskets, and sacks were filled with dried foods and stored out of the way in Cahuilla houses. Dried meat and herbs were suspended from the ceiling.

Modern:
The modern forager may be recommended to have similar storage solutions. Large storage containers that seal well can be used to store large amounts of foraged foods such as mesquite pods. Jars and bags are also useful.

I preserve the vast majority of the foods I forage by drying them. It is much easier than canning, takes up less space than freezing, and requires no electricity or special materials. Dried foods should be stored sealed away from infestation, as many insects will go after these foods otherwise. Plastic bags work alright, but many insects (and rodents) can chew right through these. Sealed glass jars are ideal. One does not need special canning jars, and any hard material with an air-tight seal will be sufficient. I keep glass, metal, and plastic containers that I would otherwise recycle and use these for dried food storage.

One should be aware that many wild food products often harbor insects such as beetle and moths. The larvae are generally edible, and can be disregarded in processing, especially if the food is cooked. However, if the foods are to be stored long-term, it may be recommended to kill these larvae by baking at the lowest setting. Enough heat should be applied for a long enough time for each batch to destroy any insect larvae or eggs. For example, one might fill a tray full of nuts and bake it at 200° F for about an hour. After cooling, one can directly transfer the nuts to a sealed container. Such a process prevents one's house from filling up with little moths and beetles that have emerged and may potentially re-infest unsealed stores.

Dried foods in hard, airtight vessels can be stored at room temperature, and it is possible that they do not degrade for multiple years.

This does carry risk of fungal or mold growth, which can be harmful to the health, so I cannot recommend it. I have stored mesquite pods, acorns, walnuts, pecans, and more, sealed in storage containers away from any moisture intrusion, outdoors for multiple years without any visible decay except for the emergence and re-infestation of insects. In some cases, moisture has gotten into the containers and caused them all to mold, rendering them unusable, so keeping the food completely dry at all times is essential.

A great way to preserve moist fruits such as plums is to make fruit leather. After cooking down, pulping and separating out the seeds and skins with a wire mesh strainer, I spread out the pulp on an oiled baking tray, then bake it on the lowest setting in the oven until it dries. I scrape it off the sheet, cut it into strips, and roll these up for an amazing food.

Sterile canning technique is also an excellent option. I use this technique for any food I want to remain moist, such as Texas persimmon pulp, wild plum preserves, or grape juice.

Freezing is also a possibility, but I never had enough freezer space to use this extensively.

Mustang grapes being spread out to dry on a platform made of juniper posts and beams, grapevine lashings, and giant cane for the surface. Note that grapes are best dried off the stem, and should be kept covered. Overlay: canned mustang grape juice.

GLOSSARY OF TERMS

In addition to the terms defined in the Methods chapters, this book uses botanical, fungal, and taxonomic terms in a particular manner, which will help the reader accurately identify the parts used and understand naming schema.

Plant terms:

Proper botanical terms will be helpful in determining exactly which plant parts were used. This book uses a combination of common and scientific terms, which are detailed below. The anatomically correct botanical terms are not always used. Terms most familiar to the layperson are preferred. For example, an underground stem may be referred to as a "root." Terms used are in bold.

Roots: underground structures. Specific varieties of "roots" may be further described by the following terms: **Bulb** – a subterranean stem with fleshy overlapping leaf bases around a small basal stem (e.g. onions). **Corm** – a subterranean stem that is aligned upright with the stem, is hard or fleshy, enlarged, and has dry scaly leaf remnants visible (e.g. water chestnuts). **Rhizome** – a subterranean stem that is horizontal, enlarged, and has scaly leaf remnants visible (e.g. ginger). **Tuber** – a subterranean stem that is an enlarged fleshy tip of a stem (e.g. Irish potatoes). **Rootstocks** – a non-anatomical term referring to certain underground parts of some species which are expanded and rich in starches and are used to store energy. These can be any of the above.

Aboveground parts: the parts of the plant above the level of its substrate. This usually includes all leaves, foliage, stems, shoots, branches, trunks, flowers, and fruits.

Fruits / cones: includes bean pods, acorns, hard seed cases, fleshy fruits, dry fruits, and all productions of the fertilized flower. The equivalent for conifers is cones, which do not originate from flowers, but from strobili (or cones). Fruits include seeds. **Seeds** – the innermost kernels of fruits that can grow into new plants. The fruit wall, or **pericarp**, surrounds the seed (or kernel). It consists of three layers: exocarp, mesocarp, and endocarp. The **exocarp** is the outer layer (skin). The **mesocarp** is between the exocarp and endocarp. In some species (e.g. plum, honey mesquite), it is fleshy or spongy and rich in carbohydrates to attract animals that eat it and disperse the seed. The **endocarp** is the inner layer that surrounds the seed. It often forms a hard shell or stone around the seed.

Flowers: includes the calyx (reduced, often leaf-like petals at the base of flowers) and every part of the flower. See diagram on p. 303.

Leaves / needles / fronds: the terminal photosynthetic structures. These include compound leaves, in which leaflets (foliolules) may be mistaken for leaves. These include petioles (leaf stems) and petiolules (leaflet stems). For conifers (Gymnospermae), these are called "needles." For ferns (Pteridophyta) and palms (Arecaceae) these are called "fronds."

Foliage: leaves and smaller stems, especially of herbaceous species.

Twigs: smaller stems of woody species, perhaps with buds or leaves included.

Branches: the foliage plus larger stems, especially of woody species.

Stems: relatively larger supporting structures of a plant, consisting of vascular and structural tissue.

Shoots: relatively young growth of stems, usually referring to straight, unbranched, green stems.

Thorns: spiny external structures. Includes true thorns (derived anatomically from stems, e.g. in mesquite), spines (derived anatomically from leaves, e.g. in cacti), and prickles (derived anatomically from epidermis e.g. in blackberries).

Bark: both inner and outer bark. It is also used when the source material does not clearly distinguish the use of outer or inner bark.

Outer bark: the outermost layer of bark up to the inner bark layer. It is usually corky and tough in woody species.

Inner bark: the phloem tissue (transports carbohydrates), which lies between the outer bark and sapwood. It becomes outer bark as it dies and a new layer of inner bark is grown.

Wood: both the sapwood and heartwood. Also used when the source material does not clearly distinguish the use of sapwood or heartwood.

Sapwood: the xylem tissue (transports water) or living wood. This is usually a lighter color than the heartwood.

Heartwood: dead xylem tissue in the center of a cross-section that is the main structural support of a tree.

Sap: the fluid transported by phloem tissue / inner bark. **Resin –** sap that has bled from the plant and dried to some degree. Conifer resin is an example. **Gum –** sap that has bled from the plant and dried to some degree. This term is typically applied to non-coniferous species, especially when the texture is gummy. **Latex** – viscous sap with a milky or opaque color.

    Juice: the liquid product of squeezing or expressing plant material.

Fungi terms:
    Fruiting body: refers to the aboveground "mushroom." Most of the fungal body (mycelium) is hidden underground, in wood, etc. The familiar "mushroom" is analagous to a plant's fruit, serving to disperse offspring in the form of spores rather than seeds.
    Mycelium: the main body of the fungus, usually comprised of a network of hyphae (fungal filaments) intermixed with a substrate such as soil or dead wood.
    Spores: the microscopic particles / powder released from fruiting bodies.

Taxonomic terms (genus / species):
    Scientific names: the most current Linnean binomial (genus and specific epithet) accepted by scientific consensus. The World Flora Online database and Kew's Plants of the World Online were used as the ultimate arbiters in cases where there was disagreement in other authoritative databases and sources. Scientific names are usually derived from Latin or Greek roots, and often refer to defining characteristics of the species. Sometimes, the specific epithet (the second part of the name) is derived from a person's name, such as a discovering botanist.
    Synonym: Linnean binomials applied to species in the scientific literature that are no longer considered current. Scientific names change as taxonomic research progresses and the taxonomy of a species and its relatives change. Scientific names are based on relatedness. Consider two superficially similar species that were thought to be in the same genus. If new genetic analyses show one of those species to be more closely related to a different genus, a change in its scientific name may be necessary, in which case the old one will become a synonym of the new one.
    Common name: non-scientific names, usually in English, applied commonly in casual language to the species. Common names can be shared by various species, and often there are many common names for a given species. The first common name given here is usually that applied by the USDA Plants Database. Common names in Spanish, French, or other languages are given in italics.

PALEOLITHIC PLANT FOODS

Premise:
    This section is a hypothetical reconstruction of the general uses of wild plant foods by pre-contact Indigenous peoples of Austin and Central Texas, who were technologically in the Paleolithic. It is based upon my research on the historical ethnobotany of Indigenous North Americans. I have spent many years in this work and can make very plausible educated guesses. It is also based on my personal experience as a native Austinite, lifetime forager, and Paleolithic technology practitioner.

    It would be far more challenging, if not practically impossible, to describe the plant component of pre-contact diets of Indigenous peoples throughout all of Texas, as cultures varied drastically between ecoregions. In East Texas, for example, the Caddo were the westernmost Mississippian peoples, a highly-developed Neolithic culture of the Southeast connected to thousands of years of agricultural development. Plant foods in West Texas were likely dominated by the asparagus family, especially *Agave*. Diets in North Texas, up to the high plains, as well as in South Texas, down to Mexico, were likely not drastically dissimilar to diets in Central Texas, but certainly varied significantly. The High Plains of the Panhandle had peoples that likely relied more on geophytes (edible underground plant parts) and animal foods.

    Natives of Central Texas in 1000 AD were small, mobile, kin-related bands of hunter-gatherers seasonally migrating within small ranges to exploit wild resources (Foster 2012). Beyond thorough treatment of agricultural phases and phrases such as "harvested wild berries, nuts, grapes, plums, and seedy plants" and "plant products roasted or baked in earth ovens contributed substantially to the diet" (Foster 2012), very little about specific wild plant foodways is discussed in scholarly works about these peoples.

    The cultural identity of these peoples is only speculative. The Tonkawa (Tickanwatic) inhabited Austin and Central Texas in the historical period. The Lipan Apache (Lépai-Ndé) inhabited Central Texas more to the Southwest. However, the Tonkawa likely migrated to Texas from the High Plains during the early contact period, and the Lipan Apache may have migrated to Texas a few hundred years before first contact. The Comanche (Nʉmʉnʉʉ) were nomadic peoples that ventured into Central Texas and Austin, but only reached Texas after the introduction of the horse by the Spanish, previously being inhabitants of the northern Great Plains. So, the cultural identity of pre-contact peoples who lived in what is now Austin is not known.

As evidenced by archaeological studies at the Gault Site and Buttermilk Creek Complex as well as Spring Lake in San Marcos, the edge where the Balcones Escarpment and Blackland Prarie ecoregion meet is one of the earliest inhabited areas known in North America and one of the longest continuously-inhabited. This area is extremely rich in natural resources, including wild plant foods such as pecan, mesquite, prickly pear, persimmon, and many others in this book. Because of this species richness, the spring waters, the high-quality chert for edged tools, its comfortably warm climate, and the meeting of various ecoregions and associated flora, the ecosystem in what is now Austin was likely highly valued by pre-contact foraging societies.

Overview:

The headlines below are the main categories of plant food use. The most important wild plant foods for carbohydrates were mesquite pods, various rootstocks, fleshy fruits, and grains. Nuts, acorns, and small seeds were important sources of protein and fat as well as appreciable sources of carbohydrates in the case of acorns. Many edible greens were essential for micronutrients. Certain species, such as yaupon and chiltepín, were highly valued for their effects or taste. Many species were harvested only opportunistically, out of necessity, or to add variety to the diet.

Animal foods likely made up around half or more of the diet. Hunter-gatherers worldwide generally had diets with an animal component ranging from 45 to 65% (Cordain et al. 2000). These diets were generally high in protein, as many wild foods are low in carbohydrates (Cordain et al. 2000). This fact, combined with the fact that mesquite pods are 85% carbohydrate and are easy to harvest and process, underscores the probable historical importance of this plant food in the Paleolithic diet of peoples in Central Texas and the Southwest.

Animal foods that were most prominent in the diets of Natives in Central Texas around 1000 AD were whitetail deer and cottontail rabbits (Foster 2012). Bison were not present in the area at the time, being restricted to more northern areas of the Great Plains, only arriving in the area in the fourteenth century when the climate cooled (Foster 2012).

Seeds & grains:
      The seeds of many herbaceous or shrubby plants, including many in the amaranth (*Amaranthus*, *Chenopodium*) and composite (*Artemisia*, *Cirsium*, *Helianthus*, *Iva*, *Solidago*, *Verbesina*) families, and the grains of certain grasses (see "Grass" in species accounts) were gathered in great abundance by certain methods. They were processed to remove extraneous matter and were usually cooked and ground into meal. These dry meals were often combined together and could be stored in large quantities for winter use. These were an good sources of macronutrients. No single species was exceptionally important, but as a general category, they represented a key food source.

Nuts:
      Nuts from pecan and, to a lesser extent, walnut trees were among the most important foods. These were unique compared to other wild plant foods in being rich in fats. They were also good sources of protein. Oil was often extracted from the nut meal and used for cooking and many material uses. Nuts were gathered and stored in great abundance and were a valuable trade good.

Mesquite pods:
      The mesquite pods, but not the beans, were one of the most important wild plant foods and were likely the primary source of carbohydrates. The pods grew in great abundance and were simple to process, requiring no cooking, shelling, or anything besides being pounded up and soaked to make a sweet beverage, or the meal being sifted to remove the beans and fibers. The meal was formed into large cakes that were stored in abundance and also used for trade.

Acorns:
      Acorns were likely not as important in the area as in other regions of North America because mesquite pods were an easier source of carbohydrates. The local oak acorns tend to be small and bitter. The Texas red oak was likely the preferred species because of its commonality and larger acorn kernel size. They were most likely eaten and were important when other wild foods had poor crop years. They were processed by various methods, but mostly by grinding them into a fine meal, leaching the bitterness out with water, and cooking it into mush or solid cakes that could be stored.

Fleshy fruits:
There were many fleshy and sweet fruits that were highly valued for their taste and as sources of sugars, vitamins, and carbohydrates. They were gathered and eaten in large quantities. Roughly in order of season, from March to January, the most important ones were dewberry, skunkbush sumac, mulberry, stretchberry, agarita, lotebush, wild plum, anacua, elderberry, wild grape, wild cherry, prickly pear, Texas persimmon, gum bumelia, flameleaf sumac, wax mallow, sugarberry, and juniper. For storage, these were either processed to isolate the pulp, which was dried, or the fruits were dried whole.

Dry fruits & beans:
There were various other fruits that could be consumed raw or cooked, including beans and bean pods. Young milkweed fruits were cooked, paloverde beans were eaten raw when green or cooked when mature, and numerous other minor species, such as catclaw acacia, were sometimes eaten.

Greens & shoots:
The leaves, stems, and shoots of many herbaceous and shrubby plants, such as those in the amaranth, mallow, and mustard families, were commonly eaten. These include nopales, or the young stem sections of prickly pears. Small clover-like herbs in the legume family were especially sought, being higher in protein. Plants such as woodsorrel, evening primrose, buckwheat, dock, greenbrier, nettle, dayflower, & spiderwort were also favored for their tasty foliage. Almost all foliage was preferred when young, tender, and soft, which was the main stage in which it was eaten raw. But greens were most often cooked, especially being added to soups and stews to improve their nutrition and flavor.

Bulbs, corms, rhizomes, & tubers:
Many underground rootstocks, where herbaceous plants store their carbohydrate reserves, were sought. They were eaten raw, cooked, or processed into dry meal or cakes for storage. Certain aquatic plants, such as southern cattail and water lily, were excellent sources of rootstocks. Numerous bulbs (e.g. *Allium*), corms (e.g. *Liatris*), and tubers (e.g. *Solanum*) were common in the prairie habitats. Wild onion bulbs were eaten regularly as a tasty and nutritious addition to stews. Many minor species with rootstocks were known and exploited opportunistically.

Bark, sap, & sugars:
 The inner bark is where sugar-rich sap flows, and the bark and sap of certain species may have been eaten for these sugars. It could be obtained in several ways. The sap of boxelder and grapevines was tapped in a way similar to sugar maples. This fresh liquid was drunk or may have been cooked down into a sweeter substance. The dried sap, called resin or gum, that exuded naturally from wounds was taken from trees such as juniper and mesquite, soaked in water, and drunk as a beverage, or simply chewed and eaten. For other trees, such as cottonwood, sections of the bark were cut out, the inner bark was scraped off, and this was boiled to extract the sugars.

 Another unique source of sugar was "honeydew"—the droppings of aphids. These bugs feed on plant sap and excrete sugary droppings that dry into a hard sugar coating on the plants, such as cottonwood leaves or reed stems. These sugar coatings on plants were dried, threshed, and winnowed to isolate the honeydew, which was a prized treat.

Flowers & pollen:
 These constituted rather minor food resources, but the flowers of many species were eaten raw or cooked. Cattail (*Typha*) pollen was gathered in surprising abundance and eaten.

Agave, sotol, & yucca:
 These plants in the asparagus family constituted significant food sources. In late winter and spring, the young flowering stalks were gathered in abundance and cooked for eating. The "heart" or inner core of agave was cut out and baked in earth ovens to produce large amounts of carbohydrate-rich food. These cooked hearts could be sliced and dried for storage. Sotol hearts were processed in a more intensive, but similar manner. The related beargrass may have also been eaten. These foods were more common in the western part of Central Texas, and may have been traded to those in the eastern part for, say, pecans.

Mushrooms:
 Few mushrooms were eaten in this arid region, but when flushes occurred, oyster mushrooms and puffballs were likely eaten and perhaps dried for storage.

Lichens:
It is unlikely any lichen species was eaten, but it may be possible to render some species edible in times of great necessity.

Beverages:
The most popular beverages were mesquite drink, yaupon tea, and sumacade. Mesquite drink was made from soaking and straining pounded mesquite pods. It was drunk for its taste and nutrition, and was lightly fermented and kept as a household staple. A strong decoction of yaupon was drunk regularly for its stimulant effects and huge amounts were drunk at certain events. Sumac fruits were crushed and soaked in water for a tart and lightly sweet drink (sumacade) perfect for hot weather.

Other beverages consumed for their taste or effects were made from various plants such as spicebush, jointfir, golden tickseed, goldenrod, greenthread, lemon beebalm, false pennyroyal, yarrow, and elderberry flowers. The seed meal of various species was also eaten mixed with enough water to make a beverage. Fresh or dried fruits and other plant foods were also commonly made into beverages.

Seasoning:
Saltbush and inland saltgrass foliage were used for a salty flavor. White sagebrush, chiltepín, wild onions, wild mustards, Mexican tea, sumac, lemonscent, wild bergamot, lemon beebalm, milkweed, gumweed, and various fruits were used for flavoring meat and in other cooking.

Gum:
Chewing gum was common, and the dried sap of various species was used plain, lightly processed, or mixed with other materials to make it. Other plant parts, such as stems, bark, buds, or fruits, were sometimes used. Species used include gum bumelia, juniper, mesquite, Indian hemp, milkweed, goosefoot, wild lettuce, wirelettuce, prickly pear, snow-on-the-mountain, compassplant, ragwort, spiny chloracantha, globemallow, cottonwood, slippery elm, and cattail.

Famine foods:
Certain difficult-to-obtain parts or poor-tasting species were used in times of scarcity. These included mature prickly pear stems, juniper inner bark, some acorn species, palmetto hearts, and bush morning glory roots. Grass grains are all edible, but most species do not produce grains

large enough to be worth the processing effort except in times of famine. The small seeds inside the beans (inedible endocarp shell) of honey mesquite pods would likely be in this category. They are laborious to extract and grind and must be cooked to destroy antinutritional compounds.

Agriculture:

There may have been no species cultivated in pre-contact Central Texas. Wild plants that were particularly useful were likely tended to some extent, at minimum by weeding or burning of habitats. Tobacco may have been planted and weeded. Bottle gourds may also have been planted for containers. Corn, beans, and squash would have been the first choices for food horticulture.

Corn, and perhaps other food crops, may have been obtained by trade with peoples in East Texas. Reasons agriculture was likely not practiced in Central Texas include the poorer soils, the aridity, presence of sufficient animal foods in the form of deer, rabbits, and bison herds (in some periods), regular burning, and the lack of necessity given the plenitude of wild plant foods.

Salt & minerals:

It is possible that salt was not used at all. If used, salt was most likely obtained from trade or certain environmental features. Salt can be gathered from particular creek banks or bottoms and saline water bodies (e.g. La Sal del Rey in South Texas), but such sources may not have occurred in Central Texas. Coastal and South Texas Natives had relatively easy access to salt.

Clay was used in some cooking or processing methods, especially to remove bitter compounds from food such as mesquite pods and acorns. The ashes of some plants may have been used for a salty addition to the diet. Wood ashes may also have been used in some food processing methods, such as removing acorn tannins.

Water:

In emergencies, water can be obtained from prickly pear, other cacti, and agave. Prickly pear can also be used to clear silt from water. Water was mostly obtained from aquifer springs, which were very common in the area. The water obtained from brooks, creeks, streams, and rivers was also generally safe to drink.

Conclusions:

The plant component of the diet of pre-contact Indigenous peoples of Central Texas was about equal in importance to the animal component. As far as simple calories, mequite pods were likely the main source. Pecans may have been a close second, especially given the caloric density of their oils. Both mesquite and pecan were likely gathered in abundance and kept stored in quantities sufficient to last multiple years.

As far as appreciated and celebrated plant foods, a handful of fleshy fruits likely featured prominently. Agarita, Chickasaw plum, mustang grape, prickly pear, and Texas persimmon were especially valued, as they were produced in abundance, were easily harvested, required little to no processing, and had delicious tastes. (These are also among the top foraging targets in this book.)

From mid- to late winter and early spring, when few fruits, seeds, or nuts were available, various vegetables and greens were appreciated. These include the flowering stalks and hearts of plants in the asparagus family such as yucca, sotol, beargrass, and agave. Nopales were also eaten at this time, as were the young greens from countless herbaceous plant species. The rootstocks of aquatic plants such as cattail, bulrush, and lotus were likely also valued as carbohydrate sources during winter.

Daily foods appreciated year-round likely included beverages made from mesquite pods, sumac fruits, and yaupon. Mesquite pod meal cakes and pecan oil may have also been daily staples. Wild onions were likely commonly cooked with meat dishes, and various plants such as chiltepín and lemon beebalm were used as spices in such dishes.

Foods were often cooked in earth ovens. Large caches of dried foods were always kept in anticipation of poor seasons and harvests.

Accompanied by regular fresh deer, rabbit, and other meats, the diet of people in Central Texas a thousand years ago was incredibly rich and nutritious, with a remarkable diversity that changed with the seasons, and few times when appreciable wild plant foods were not available.

SPECIES ACCOUNTS

Overview

The "History" sections detail historical Indigenous uses of plants for food. The names of the peoples to which specific uses are ascribed are listed in a reference table in the index. Generally, the peoples referred to are in one or more publications. These include ethnographies, ethnobotanical studies, historical accounts, and other sources, which are often primary. These historical and ethnographic references may be cited elsewhere with the author name and year.

The "Notes" sections may include information about the general character of species, their season, their nutritional content, and my personal practices foraging them. Phytochemical, botanical, and other sources are in a separate reference chapter and are cited by superscript numerals.

Most accounts cover a particular species and sometimes include information about how other species in the same genus were used. Some accounts summarize all the uses of a particular genus that is found in Texas. This is done when many similar species in that genus were used in a similar way, informing the possible uses of the local species, for which the historical data may be lacking. When the account covers a genus, it sometimes discusses plant species not found in Texas. Texas species in the genus that are found in Travis County and are not rare are listed in each genus account.

QR codes are used throughout the book. These link to video playlists on YouTube.com, all made by me on my channel, Paleo Foraging. Some of these link to all videos on a particular topic (e.g. "mushrooms"), while others link to all videos I have made about a particular species (e.g. "mesquite"). The QR codes are labeled by the text they are directly under or directly beside.

By scanning a QR code with a smartphone camera, the reader can reveal a link which can then be clicked to be directed to a YouTube playlist. These videos may help with visual identification of species and demonstrate practices such as preparing mesquite beans for eating. In these videos, I also often cover many non-food uses of species which are not covered in this book.

FUNGI

LYCOPERDACEAE – Puffball Family

Lycoperdon spp.
Puffball

Cayuga: *duwatagehänegqus* – "smoke shoots out," Mohawk: *o'tgu raona'daro* – "devil's bread," Onondaga: *dewadi`ɛ'gwae'gwas* onä´'sa' – "smoking fungus" / *deyutwi'no'ni's unä'sa'* – "round fungus," Haudenosaunee: *onĕⁿ'să'wa'nĕ'*, Kiowa: *ai-peep-o-pä*
Loc.: E & C TX; not uncommon in Travis Co.

History

Fruiting body – The edible young stage was eaten after cooking (Haudenosaunee, Kiowa, Zuñi). They were either baked (Kiowa) or peeled, sliced, and fried in grease or sunflower oil (Haudenosaunee). They were gathered in large quantities and dried for storage (Zuñi).

Notes

Character – The two species below (and puffballs in general) had similar uses, with both being historically used as described. Puffballs are inedible once they reach the growth stage in which they begin forming spores. This can be judged by their firmness and by cutting them in half. Edible puffballs are firm, white, and homogeneous throughout, whereas the inedible stage begins when they are darker in their centers but still solid.

L. perlatum is the most common in Texas and its outer surface is covered with many tiny conelike warts. *L. pyriforme* is also common, but has a smoother outer surface with scaly patches, and its base is more tapered. *L. marginatum* is also in Travis County and is identifiable by its rough outer 'skin,' which forms a distinct layer.

Calvatia is another genus in the puffball family that is edible, also only in its young, homogeneous flesh stage. These are much larger but are less common in Travis County than *Lycoperdon* puffballs, although *Calvatia craniiformis* (brain puffball) is not uncommon in Travis County.

Season – Puffballs can be found year-round, but are most common in cooler weather (late fall through early spring) after rains.

Practice – I find puffballs (*Lycoperdon*) in Central Texas primarily in lawn-like habitats or on the forest floor, presumably with some underground decaying matter supplying their growth. They flush year after year from the same spots. They tend to be pretty small, but if I encounter a large flush of them, it has been worth my efforts to harvest some. I make sure they are young and firm. Simply tapping on them and feeling a solid, bouncy, "thunk" generally signifies they are firm and still edible. Like all mushrooms, they should be cooked thoroughly before consumption. They taste great sliced and sautéed in a bit of pecan oil.

Lycoperdon perlatum Pers.
Common puffball

= Lycoperdon gemmatum, L. bonordenii, Calvatia cyathaformis, Bovista plumbea
Dakota: *hokshi chekpa* – "baby's navel," Pawnee: *kaho rahik*
Loc.: E, NE, SE, & C TX; not uncommon in Travis Co.

Lycoperdon pyriforme Schaeff.
Pear-shaped puffball

= Apioperdon pyriforme
Cherokee: *nàkwisiusdí / nɔ.kʷi.sju.ti.gi'do*, Menominee: *iniki'wi opa'skûk*
Loc.: E & sparse C & SE TX; not uncommon in Travis Co.

Additional notes:

Puffball spores, including those of *L. perlatum* and *L. pyriforme*, were used medicinally by Indigenous peoples. The dry or moistened spores were used to stop bleeding or to aid the healing of sores (Cherokee, Northern Cheyenne, Dakota, Kiowa, Omaha, Ponca, Round Valley Natives). They were also used as a baby powder, to prevent chafing and rashes under the armpits or between the legs of babies (Northern Cheyenne, Menominee). The spores were applied to the cut umbilicus of newborn babies (Dakota, Kiowa, Omaha, Ponca), thus the Dakota name for *L. perlatum*. The spores of mushrooms, likely puffballs, were applied to tooth cavities to treat toothache (Comanche).

Puffballs mature into hollow balls full of spores, making them an ideal source and container for spores, which fruiting bodies of other mushrooms often waft into the air from gills or wide surfaces. Extracts of puffball mushrooms were found to be significantly active against gram positive and gram negative bacteria as well as fungi.[23]

Lycoperdon perlatum. Base photo: young fruiting body on the ground in late October. Overlay, clockwise from top left: range map of *Lycoperdon* spp. (GBIF); close-up of young fruiting body; harvested and cleaned puffballs; puffball broken open, revealing homogeneous white flesh, indicating they are suitable to eat.

PLEUROTACEAE – Oyster Mushroom Family

Pleurotus ostreatus (Jacq. e*x* Fr.) P. Kumm.
Oyster mushroom
Loc.: E, C, SE, & NE TX; common in Travis Co.

History
Fruiting body – Oyster mushrooms were eaten by Natives in the Pacific Northwest.

Mushrooms found on dead limbs of *Populus* spp, *Platanus* spp., *Salix* spp., and *Quercus* spp. were a popular food of the Cahuilla. They were gathered in the spring when the mushrooms were whitish and the sap had begun to run. They were still considered edible later in the year when turned brown. They were boiled, fried, used in making gravy, or mixed with acorn mush. These may have been, or have included, oyster mushrooms, which match the description.

Notes
Character – These mushrooms are found on dead and decayed parts of living trees, especially on willow (*Salix* spp.) and cottonwood (*Populus deltoides*). Standing dead branches that are more than three inches in diameter are a particularly favored habitat of oyster mushrooms. Oyster mushrooms are characterized by colors ranging from whitish to cream or light brown, a shelf-like form, soft, flexible tissue, and gills that extend from the cap margin all the way down to their point of emergence from the wood.

In some areas of Texas and North America, huge amounts of oyster mushrooms, weighing dozens of pounds, can sometimes be gathered from a single colony. I have never seen a colony in Central Texas that weighed more than a couple of pounds.

Season – After rains, especially in the cooler weather (late fall through early spring), flushes of oyster mushrooms can be found that are worth harvesting.

Practice – I have gathered them from dead branches of black willow (*Salix nigra*), oak (*Quercus* spp.), and Chinese tallow tree (*Triadica sebifera*). Like most mushrooms, I prefer them sliced and sautéed in a bit of oil. They have a meaty texture and savory flavor. The oyster mushrooms in Central Texas tend to be on the small side, and mushrooms in general are not a reliable or consistent food source in this rather arid region.

Pleurotus fruiting bodies. Note that the gills go down to the point of emergence. Both are on dead black willow (*Salix nigra*) limbs. Top photo in late December, bottom photo in early December. Range map (*Pleurotus* spp.): GBIF.

POLYPORACEAE – Bracket Fungi Family

Trametes versicolor (L.) Lloyd
Turkey tail

= Polystictus versicolor
Loc.: E & C TX; common in Travis Co.

History

Fruiting body – When young and tender, these were eaten by the Dakota, except those growing on ash trees (*Fraxinus* spp.), which they said are too bitter. They were prepared by boiling them.

Notes

These are rather woody and tough, but can be adequately palatable when boiled. They can be easily confused with mushrooms in the genus *Stereum*. *Trametes* has tiny pores on the white underside of the fruiting bodies, whereas *Stereum* undersides are smooth with a yellow or tan color. Turkey tail is considered a medicinal mushroom, so may have health benefits, though in my research, I did not find any historical Indigenous medicinal use for it.

Trametes versicolor. Note porous structure on underside of fruiting body seen on the right side of the left photo. Range map: GBIF. Photos: GBIF.

Additional Notes:

Another mushroom worthy of consideration for foraging in Central Texas is *Laetiporus sulphureus* (chicken of the woods).

CONIFERS

CUPRESSACEAE – Cypress Family

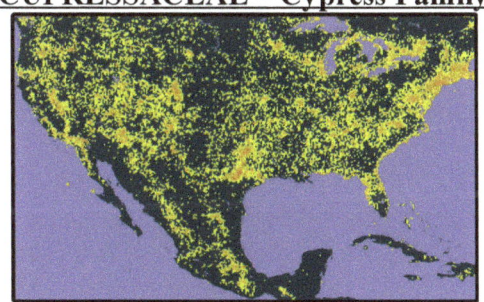

Juniperus distribution. Source: GBIF

Juniperus spp.
Juniper

Cedar
Comanche: *waapl*, Diné: *kat / tilk'yis* (when dry) – "it crackles," Gosiute: *wa'-pi*, Kiowa: *k'okiädlä*, Lakota: *xoŋdse,* Osage: *xaŋte*
Loc.: all TX except far S; very common in Travis Co. (2 spp.); 12 spp. in TX (4 cultivated).
Form: shrub, tree; perennial.

History

Cones – The cones (fleshy berries) were eaten by the Chiricahua, Mescalero, and Western Apache, Cahuilla, Comanche, Diné, Gosiute, Hopi, Isleta Pueblo, Jemez Pueblo, Kawaiisu, Kiowa, Tewa, Northern Ute, Natives in Arizona, California, and New Mexico, Western Natives, American Indians, and Mexican folk. Species eaten include *Juniperus californica*, *J. deppeana*, *J. monosperma*, *J. occidentalis*, *J. osteosperma*, *J. scopulorum*, and *J. virginiana*. They were eaten in large quantities.

Berries from various trees were sampled to find the sweetest, as trees vary in the taste of the fruits they produce, ranging from sweet to bitter (Cahuilla, Comanche, and Northern Ute). Juniper berries were gathered in the fall (Comanche) or October and November (Apache). The berries were struck from the tree and caught in a basket or were gathered from the ground (Kawaiisu).

The berries were eaten raw or cooked, and were sun-dried for storage. They were roasted on hot stones to eat whole or were ground into a meal (Apache). The berries were sometimes cooked by boiling (Apache, Isleta, Kawaiisu) or on a pan (Apache, Diné, Hopi, Tewa) with water added to make a thick sauce used like gravy (Apache). The berries were eaten with oil (Gosiute), re-hydrated dried cooked agave heart (Apache), or in corn bread or stews (Hopi).

They were rubbed on a metate with a mano to separate the seeds from the pulp, which was eaten fresh or dried (Northern Ute). They were also ground whole into a meal and made into mush or cakes (Cahuilla, Kawaiisu). The Kawaiisu dried the berries, gently pounded them and sifted out the seeds, then pounded the dried flesh into a meal that was eaten dry or was moistened, molded into cakes, and dried on leaf bases for storage. These were stored in earthen pit caches lined with brush and buried.

To improve the taste and digestibility, Natives in Arizona and New Mexico dried the berries, ground them into a meal, and mixed this with water to form cakes that were sun-dried. These dry cakes contained 14.3% water, 5.7% protein, 17.9% complex carbohydrates, 11% sugar, and 3.9% ash (Palmer 1871).

Needles – An infusion or decoction of the needle sprays was drunk as a beverage, especially in the winter (Gosiute, Jemez).

Inner bark – The inner bark of *J. monosperma* was eaten in times of scarcity (Diné).

Sap – The sap of *J. monosperma* was chewed as gum (Tewa).

Notes

Character – Though I have found no direct evidence for the historical use of *J. ashei* for food (or anything else), its closest relative is *J. monosperma*, which was historically eaten. The other *Juniperus* species in Travis County, *J. virginiana*, was eaten and gathered in quantity to dry and store for winter use (Comanche, Kiowa). *J. virginiana* is more common east of the Balcones Fault, being the dominant *Juniperus* species of the Blackland Prairie Ecoregion and eastward, whereas *J. ashei* is the most common tree species in the Edwards Plateau (Hill Country).

Season – Individual trees produce berries every two years, but the berries are available every year. In Travis County, juniper berries are found ripening from fall through winter. The needles can be used year-round.

Nutrition – Dried *J. monosperma* berries contain 4% protein, 3% fat, 43% carbohydrate, 45% fiber, potassium (646 mg/100 g), calcium (404 mg/100 g), sodium (130 mg/100 g), phosphorus (14 mg/100 g), iron (3 mg/100 g), zinc (1 mg/100 g) and copper (1 mg/100 g).[75] These are excellent sources of fiber, potassium, and calcium, and moderate sources of carbohydrate, sodium, iron, and copper.

Practice – I have eaten large amounts of *J. ashei* berries every year for decades, and they an excellent foraging target. It is crucial that one sample trees before gathering, as there is a wide range in palatability of the berries, with some trees producing dry, insipid, astringent, or bitter berries and other trees producing juicy, flavorful, and sweet berries. One easy way to gather them is with the "strike and catch" method. They can also be gathered by inserting a whole fruiting branch into a canvas tote bag and shaking it. The berries of *J. virginiana* must often be gathered from the ground, as they grow more upright.

Juniperus ashei J. Buchholz
Ashe juniper

Mountain cedar
Loc.: C & SW TX; very common in Travis Co.
Form: shrub, tree, up to 30 ft. tall; perennial.
Food: no records.

Juniperus virginiana L.
Eastern red cedar

= Sabina virginiana
Virginian juniper, red juniper, pencil cedar, Carolina cedar, red savin
Comanche: *ekawa:pᵛ (eka* – "red") / *tʉbitsiwaapī* – "real cedar" / *wapokopī* / *waapokopī* (juniper berries), Creek: *əcenə*, Dakota: *hante* / *hante sha* (*sha* – "red"), Omaha-Ponca: *maazi*, Pawnee: *tawatsaako*, Kiowa: *'ko-kee-äd-la, ahi'ñ, a-heeñ* – "peculiar" or "conspicuous," *ya-'toñ-bä* – "wood for love flute," Diné: *kat-tiltxatíh* – "juniper, crackling" / *tilk'yistchíí'* – "juniper-wood, red" / *kat·'ni'eełiih* – "juniper, soft"
Loc.: E, N, & C TX; common in Travis Co.
Form: tree, up to 90 ft, tall; perennial.
Food: Comanche, Kiowa.

Additional notes:

Junipers had many important non-food uses among Indigenous peoples of North America. The fruits and needles were used medicinally, especially an infusion or decoction being drunk for coughs, colds, and related ailments.

The needles were a common incense, which was also used medicinally (the smoke treatment), and ceremonially, being considered a purifying element. Juniper needle sprays are my favorite incense; I always have dried needle sprays available as they ignite easily and their smoke is pleasantly aromatic.

The shredded outer bark was commonly used for tinder, and it is probably the best source of such material in the area. The wood is very rot resistant, even in contact with soil or water, and it was a common material for building houses among Indigenous peoples. The houses of early settlers in Austin were made entirely of Ashe juniper wood (framework, siding, and roofing), examples of which can still be seen.

Juniperus ashei. Base photo: small tree in a remnant prairie. Overlay, clockwise from top left: male cones (strobili) in early January; range map (GBIF); shaggy bark; ripe berries in late December; berry cut in half showing thickness of flesh; harvested berries set to dry; compact needle sprays (compare to looser needle sprays of *J. virginiana*).

Juniperus virginiana. Base photo: large tree in winter (note straight, vertical trunk). Overlay: spreading needle sprays (compare to compact needle sprays of *J. ashei*); range map (GBIF).

CACTI

CACTACEAE – Cactus Family

Cylindropuntia leptocaulis (DC.) F.M.Knuth
Tasajillo

= Cylindropuntia brittonii, Grusonia leptocaulis, Opuntia brittoni, O. frutescens, O. leptocaulis
Christmas cactus, pencil cactus, turkey cactus, coyote cactus, *tasajulla, garrambulo*
Apache: *xucntsai* – "crazy cactus," Seri: *iipxö*
Loc.: all TX except E & far N; not uncommon in Travis Co.
Form: succulent, cylindrical segments, up to 5 ft. tall; perennial.
Flowers: Apr-Aug (yellow, green).

History

Fruits – These were eaten by the Apache, Seri, and Texas Natives 5,000 years ago. They were spread on the ground and carefully brushed with branches of creosote or other soft brush to remove the spines, then were eaten fresh (Seri). The fruits were also crushed and mixed with *tiswin*, a fermented beverage (Apache).

The fruits of at least nine other *Cylindropuntia* species were eaten by the Cahuilla, Hopi, Seri, Tewa, Tohono O'odham, and Zuñi. In general, the fruit's spines were rubbed off and they were eaten raw or cooked, or dried for storage. The stems and flower buds of at least six *Cylindropuntia* species were eaten, being usually cooked after spine removal. Thus, the stems and flower buds of tasajillo are likely edible, especially cooked. However, their small size, like the fruits, may not warrant the effort to process them. *C. imbricata* is not uncommon as an ornamental in Travis County and its flower buds, fruits, and stems were eaten by the Tewa and Tohono O'odham.

Notes

Character – The fruits are like miniature prickly pears. They are sweet, juicy, fruity, and delicious. Unlike prickly pears, which have many hard seeds, tasajillo seeds are small and can be eaten whole with the fruit. The fruits are covered with glochids (tiny spines), which can be brushed off in the same manner as prickly pears. The fruits often also have small green stem segments attached. These segments drop off when animals feed on the fruits and can grow into new plants (vegetative reproduction).

Season – Tasajillo fruits have a very long fruiting season, and in Central Texas, I find them year-round, depending on the individual plant. However, they mostly flower in the spring and early summer and fruit from late summer to mid-winter. Its name "Christmas cactus" refers to its fruiting in late December, which is unusual among plants, but is not the peak availability of its fruits.

Practice – To gather the fruits, I use a stick to knock them off the plant into a container. I use sticks to knock off the green segments from the ripe fruits. Then, I brush off the glochids using a wad of grass or juniper needles. I either place the fruits within a large wad of grass and work them around to brush them or hold them in place with a stick as I rub them with a handful of juniper needles. Then I eat them whole, fresh, and raw. The return on calories for time expended to process the fruits is rather low but they make a very tasty treat and are more delicious than prickly pears, in my opinion.

Cylindropuntia leptocaulis (L) and *Opuntia engelmannii* (R).
Note that both *Cylindropuntia* (cholla) and *Opuntia* (prickly pear) are in the subfamily Opuntioideae.

Cylindropuntia leptocaulis. Base photo: fruiting plant in late November. Overlay, clockwise from top left: ripe fruits with short stem segments attached; handful of harvested and de-spined fruits; close-up of ripe fruits, glochids, and stem segments; range map (GBIF).

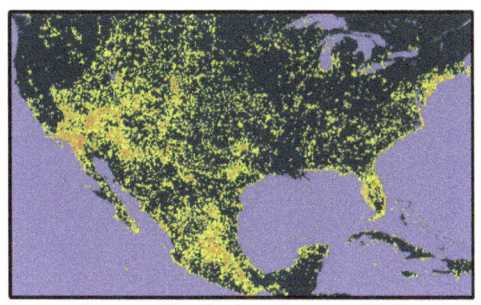 

Opuntia distribution. Source: GBIF.

***Opuntia* spp.**
Prickly pear, *nopal*

Apache: *gultice*, Comanche: *wɔkwɜsi / wokweesi / wokwéesi / wokweebi*, Diné: *xwoc* – "cactus," Hopi: *yün'yü / yüñü*, Jemez: *jiǽpooh*, Kiowa: *sen-adl-gaw / señ'äló / sen-a-lo / äló / a-lo*, Lakota: *uŋkčela blaska* – "flat cactus," Tewa: *sæ*
Loc.: all TX; very common in Travis Co.
Form: succulent; low shrub, shrub; perennial.

History

Water – *Opuntia polyacantha* stems were sometimes used to obtain water by an unspecified method (Cheyenne). *Opuntia* stems were also cut into strips and put into silty or cloudy water (sometimes after being grilled first) to cause the sediment to settle to the bottom to use the water for drinking or washing (Mexican folk, Texians). *Cylindropuntia fulgida* stems were similarly used by the Seri. The mucilaginous character of the inner stem tissue and its juices in both genera is likely responsible for this phenomenon, with the mucilage adhering to suspended particles, causing their precipitation.

Flowers – The flower buds were eaten by the Cahuilla, Kawaiisu, and Timbasha. They were broken off into baskets with sticks (Cahuilla, Timbasha) or a flattened, sharpened stick (Kawaiisu). The ends were cut off, an incision was made lengthwise, and the skin was peeled, along with the spines (Cahuilla). The buds were cooked before eating (Kawaiisu) in an earth oven for twelve or more hours (Cahuilla). Species eaten include *O. basilaris* and *O. ficus-indica*, but many species were eaten.

Opuntia flowers were eaten by the Timbasha, Tohono O'odham. They were washed, then fried in grease (Tohono O'odham).

Seeds – The seeds of all *Opuntia* species were parched, pulverized, and cooked into a gruel or mush (Cahuilla, American Indians). Seeds were threshed from the dried fruits by spreading them out and striking them with a palm frond petiole (Cahuilla). They were winnowed, further dried, ground into meal, and cooked into a mush or beverage.

Fruits – Prickly pear fruits from various species were eaten by the Apache, Cahuilla, Cenis, Cheyenne, Coahuiltecans, Comanche, Dakota, Diné, Gosiute, Hopi, Isleta, Jemez, Karankawa, Kawaiisu, Mexican Kickapoo, Kiowa, Mayo, Pawnee, Puebloans in 1540, Seri, Tewa, Timbasha, Tohono O'odham, Texas Natives 5,000 years ago, Plateau tribes of the Pacific Northwest, Natives in Southern California, other Natives, Mexican folk, and Texians. Species eaten include *Opuntia basilaris*, *O. engelmannii*, *O. erinacea*, *O. ficus-indica*, *O. gosseliniana*, *O. humifusa*, *O. phaeacantha*, *O. polyacantha*, and *O. tuna*.

Prickly pear fruits were an important food source and were eaten in large quantities (Apache, Cahuilla, Cenis, Coahuiltecans, Karankawa, Tohono O'odham, Puebloans in 1540, Texas Natives in the Archaic). They were perhaps the most important dietary fleshy fruits among Natives in the Southwest. Cabeza de Vaca, around 1530, referred to the Karankawa as "*los de los higos*" – "those of the figs," referring to their staple consumption of prickly pear fruits, which were known at the time as "Indian figs." Cabeza de Vaca spent much of his time contributing to the prickly pear harvest, which, along with mesquite, was a staple food of Texas Natives at first contact. Long trips were made by the Western Apache to gather each kind of prickly pear at the known time of ripening for each species.

To gather the fruits, wooden tongs were used, which could be a pliable branch doubled over (Apache, Hopi, American Indians). A forked stick was also used (Diné), or they were simply knocked off into a basket using a stick (Cahuilla).

A point of mesquite heartwood was attached to a long shaft of cane by Natives in Southern California to spear prickly pear fruits without getting close to the cactus spines. A small crosspiece was tied to the point to serve as a hook or to prevent the spear point from penetrating too deeply into the fruit. To make the prickly pear fruit container, mesquite bark was used to interweave cane splints, forming a cylinder, with a bottom of interwoven mesquite bark. This was carried upon the back.

To remove the tiny spines (glochids), the fruits were rubbed with a brush or bunch of grass or twigs, either on the plant or held in tongs (Apache, Cahuilla, Kiowa), rolled in sand or on the ground with a branch, or rubbed with buckskin (Apache, Northern Cheyenne, Comanche, Hopi, Jemez, Karankawa, Seri, Tohono O'odham).

The fruits were eaten fresh or dried for storage (Apache, Diné, Tohono O'odham, other Natives). They were split lengthwise, their seeds were removed, and the fruits were spread on grass (e.g., *Andropogon gerardi*) to sun-dry (Apache, Diné, other Natives). Dried fruits were eaten as is or boiled or soaked in water before eating them (Apache, Comanche, Diné, other Natives). Dried fruits were also pounded fine and mixed with fat to eat (Comanche).

To preserve the pulp, the Tohono O'odham scooped it out of the skins, removed the seeds, and spread the pulp on grass on the ground to sun-dry for two days, after which it was stored in sealed pottery jars. To make syrup, the pulp was mashed in a basket with a stick, then the juice was squeezed out, strained twice, then boiled and again strained.

Fruits were eaten fresh and whole or were made into preserves, jam, or butter (Apache, Kiowa, Tohono O'odham, Puebloans in 1540, Mexican folk). Sometimes, the fresh juice was extracted and drunk (Apache). The fruits were boiled and eaten, sometimes with cornmeal porridge added (Tewa). The fruits were sometimes fermented into a drink (Mayo).

Stems – The stem segments (pads) of various *Opuntia* species were eaten by the Western Apache, Cahuilla, Dakota, Gosiute, Hopi, Isleta, Mexican Kickapoo, Pawnee, Timbasha, Tohono O'odham, Texas Natives in the Archaic, Pacific Northwest Natives, other Natives, and Mexican folk. Species eaten include *O. basilaris*, *O. engelmannii*, *O. erinacea*, *O. ficus-indica*, *O. humifusa*, and *O. polyacantha*. Generally, only the young and tender stem segments were eaten.

The large spines were removed by scraping, burning, brushing, or rubbing with grass, with any remaining tiny spines (glochids) being washed or rubbed off (Cahuilla, Flathead, Hopi, Timbasha, Tohono O'odham). Sometimes, American Indians roasted the stems in hot ashes and when cooked, the outer skin, including the spines, could be easily peeled off and the inner part eaten. Stems were also peeled to remove the spines (Isleta, Mexican folk).

The stems were cooked by boiling (Cahuilla, Hopi, Isleta, Timbasha), in earth ovens (Timbasha, Texas Natives in the Archaic), in soup (Mexican folk), or by fire-roasting (Dakota, Gosiute, Pawnee). They were often cut into small pieces before or after cooking. Cooked stems were sometimes dipped in a syrup made from boiling baked sweet corn before eating them (Hopi).

The stems were an important food source, especially in the spring, when other foods were scarce (Western Apache, Cahuilla, Hopi). They were sun-dried for storage, sometimes being cooked in an earth oven first,

and the dried stems were prepared to eat by boiling them (Cahuilla, Timbasha).
Sap – The gum exuding from the stems was eaten by some Natives.
Roots – The roots were sometimes eaten by Plateau tribes of the Pacific Northwest.

Notes

Character – The fruits, called *"tunas"* in Spanish, are one of the most well-known wild edibles of the Southwest. The genus is native to almost every region in the Americas from Argentina to Canada, but is most abundant in deserts and other arid regions.

The flower buds look like small, green prickly pears with a floral "hat." After flowering, the petals and sepals die and fall off, leaving a circular scar on top of the fruit that lacks spines.

Season – The flower buds are available from early spring to early summer. The stems are best when they are young, tender, and flexible. At this time, they will still have the long, curved, conical, vestigial leaves that drop off when the spines mature. These young stems are found in mid-spring to mid-summer. The fruits ripen in midsummer to early fall, but can be sometimes found as late as mid-winter.

Nutrition – *Opuntia engelmannii* fruits contain 8% protein, 93 mg/100 g magnesium, and 340 mg/100 g potassium.[26] They are good sources of magnesium. Ripe *O. ficus-indica* fruits contain 2 μg/100 g vitamin A and 7 μg/100 g lycopene.[44] The dried stems of *O. dillenii* contain protein (91%), magnesium (61 mg/100 g), calcium (32 mg/100 g), vitamin B_{12} (140 mg/100 g), and vitamin A (35 mg/100 g).[4] These levels of protein, vitamin B_{12}, and vitamin A are extremely high and may reflect measurement or reporting errors.

Practice – Prickly pear glochids can be easily rubbed off, but the large spines adhere strongly to the stems, and must be scraped or burned off. The fruits only have glochids; stems have glochids and large spines.

To harvest the fruits and brace them to rub off spines, leather gloves work fairly well, but can still get spines penetrating. I sometimes use a sharpened stick to spear the fruit, twist it off, and hold it on the ground to rub off spines. The top center of the fruits (flower scar) always lacks spines, so I sometimes use this spot to brace the fruit with one finger as I rub off the spines or to knock it off the plant, but that is risky. I sometimes also simply use a nearby broken stick or branch to knock the fruit off and roll it around.

Tongs work the best, and I sometimes use a green juniper branch that I bend in half. I grasp the fruit and sometimes rub off the spines while it is still on the plant. Otherwise, I pick it off and hold it as I rub it. To rub off the spines, I usually use a wad of grass or juniper twigs. A dedicated brush (such as a skin brush or wrapped bundle of grass) works well for processing a lot. The spines can also be removed by vigorous washing, burning, or peeling, but I never use those methods because I prefer to rub off the spines before placing the fruit in my collection container. If fruits are placed together with spines still on them, the spines embed into neighboring fruits and are more difficult to remove. So, I remove the spines as soon as I pick the fruits in the field.

In my ideal gathering technique, I use cheap metal kitchen tongs in my left hand to grasp and twist off the fruit, then bring it to my right hand, where I hold a natural fiber skin brush that I use to brush off the glochids while still grasped in the tongs. I then grasp the fruit by two fingers, rotate it 90 degrees, and brush the small spots where the tongs held it, then a final sweep to ensure complete glochid removal. Then, I put it in a canvas tote bag clipped on my backpack with the other prickly pears and move to the next fruit.

To preserve the fruits, I slice them in half and dry them. The seeds can be removed before or after drying, but are much easier to cleanly remove after drying. The dried fruit halves taste like a delicious fruit leather. They can also be rehydrated by soaking and used as if fresh.

The dried seeds can be ground into a meal in a stone mortar and pestle. This can be drunk or cooked into mush. To make a sticky meal, they can also be ground up along with the rest of the dried fruit.

The flowers have a large pedicel (base) that becomes the fruit. Both before and after the flower blooms, this green pedicel has the appearance of a small, green prickly pear. I have eaten these flower buds after slow-cooking them in a manner similar to the above method used by the Cahuilla, and they make an excellent vegetable. I have found they can be harvested before or after flowering and are indistinguishable in taste and consistency.

WARNING – The main dangers of prickly pear foraging are the spines and a reaction when eating them the first time. The tiny glochid spines are dangerous to consume. A member of the La Salle expedition to Texas died from not removing these before eating the fruit.

Historical accounts (Apache, Texians, Tohono O'odham, Spanish expeditions) refer to an illness seemingly caused by eating the fruits for the first time, or eating them in excessive quantities. This can cause fever, chills, headache, nausea, diarrhea, and delirium lasting for a day or so. I have known this to happen in people eating the fruits for the first time. The cause is difficult to ascertain, but it may be some phytochemical property of the plant or a consequence of its associated microbes. The Tohono O'odham said that this only happens with one of the two species they eat. After the illness, one is thought to become acclimated and no longer fall ill.

Opuntia engelmannii Salm-Dyck
Cactus apple

= Opuntia lindheimeri (*O. engelmannii* subsp. *lindheimeri*)
Engelmann's pricklypear
Cahuilla: *qexe'yily*, Tohono O'odham: *nohwi*
Loc.: all TX except N & far E; very common in Travis Co.
Form: succulent, flattened segments, up to 5 ft. tall; perennial.
Flowers: Apr-June (yellow, pink).
Food (fruits and stems): Cahuilla, Mexican Kickapoo, Tohono O'odham, other Natives, Mexican folk.

Opuntia ficus-indica (L.) Mill.
Indian fig

= Opuntia megacantha, Cactus ficus-indica
Barbary fig, *nopal*
Cahuilla: *navet*, Seri: *heel cooxp* – "white *heel*"
Loc.: C & SW TX; common ornamental in Travis Co.; introduced.
Form: succulent, flattened segments; perennial.
Notes – The specific epithet "*ficus-indica*" is Latin for "Indian fig." This species was likely domesticated in Mesoamerica and is currently cultivated worldwide. The large spines are almost non-existent, and there are fewer glochids (tiny spines) than other *Opuntia* species. It is grown commercially for its fruits (*tunas*) and stems (*nopales*), and as a host of cochineal scale insects (*Dactylopius* spp.) which are a major food dye source. Most plant databases do not record this species as occurring in Travis Co. or Texas (BONAP, iNaturalist, POWO, USDA, WFO). However, this species is a common landscaping plant in Texas, especially in Austin, where it is also sometimes found growing wild.
Food (fruits and stems): Cahuilla, Mexican folk.

Opuntia humifusa (Raf.) Raf.
Devil's tongue

= Opuntia austrina, O. compressa
Low prickly pear
Dakota: *uⁿchela*, Pawnee: *pidahatus*
Loc.: E, S, & C TX; not uncommon in Travis Co.
Form: succulent, flattened segments, up to 18 in. tall; perennial.
Flowers: Feb-Aug (yellow).
Food (fruits and stems): Dakota, Pawnee.

Opuntia engelmannii. Base photo: patch of plants with ripe fruits in late August. Overlay, clockwise from top left: ripe fruit on plant in mid-August (note each dot is full of tiny glochid spines); harvested, de-spined fruits; young stems (note leaf remnants) ready for harvesting in mid-April; flower buds in early April; range map (GBIF); ripe fruit cut in half in early December.

TREES AND SHRUBS

ACERACEAE – Maple Family

Acer negundo L.
Boxelder
= Negundo aceroides, N. negundo, N. interius, Rulac negundo
Box elder maple, ashleaf maple, three-leaf maple, red river maple, *fresno de guajuco*
Apache: *tciłntsei* – "large ash," Northern Cheyenne: (the sap) *me?eshkehama?*, Dakota: *tashkada^n*, Teton Dakota: *cha^n-shushka*, Gosiute: *gu'-su-wup*, Diné: *sool*, Hocąk: *nąhošge* – "tree grows fast," Kiowa: *kaw-señ-añ-daw*, Meskwaki: *mämänenatähonitōtomic*, Ojibwe: *adjagobi' mûk*, Omaha-Ponca: *zhaba-ta-zho^n* – "beaver wood," Pawnee: *ósako*, Tewa: *te'jidi*, Winnebago: *nahosh*
Loc.: E, C, & SE TX; very common in Travis Co.
Form: tree; up to 50 ft. tall; perennial.

History

Inner bark – Sections of bark were removed by the Chiricahua and Mescalero Apache, and the layer of inner bark was scraped off, boiled, and reduced to extract its sugars. These scrapings were dried and stored for winter.

The Haudenosaunee dried the inner bark of *Acer saccharum*, pounded it in a mortar, sifted out the fibers, and made it into a kind of bread.

Sap – Boxelder trunks were tapped like a maple tree and boiled to form a sugar (Northern Cheyenne, Kiowa, Dakota, Omaha, Ponca, Winnebago, Pawnee). The fresh sap was also drunk as a beverage (Ojibwe). The sap was drained into a pouch formed from a young deer stomach turned inside-out (Northern Cheyenne). Once full, the pouch was tied shut and hung up to store. Scrapings of the insides of hides (for collagen / gelatin) were added to the boiling syrup to make a candy. The sap was added to tea to sweeten it. Boxelders were tapped in the early spring (Northern Cheyenne).

The sap of *A. glabrum, A. negundo, A. rubrum, A saccharum*, and *A. saccharinum* was gathered by Indigenous peoples in North America to boil down into sugar or drink fresh (Chiricahua and Mescalero Apache, Santee Dakota, Haudenosaunee, Hocąk, Menominee, Meskwaki, Ojibwe, Forest Potawatomi).

Tapping sugar maple was historically achieved by making a diagonal cut in a tree about 3.5 inches long about 3 feet from the ground (Ojibwe). At the low end of this cut, the bark was removed in a vertical line, 4 inches downward. At the bottom of this, a wooden spigot that measured about 6 inches long and 2 inches wide was driven, being inserted into a cut made for it. The sap dripped out of this and was collected in containers below. The Haudenosaunee tapped maple trees for sap by sawing a slanting gash into the trunk with a chert knife, then driving a flat stick into the gash to direct the dripping sap into bark receptacles below. The cut was 2 inches deep and 10-12 inches long, with the flat stick at the end of the sloping gash.

Notes

Character – *Acer saccharum* (sugar maple) is the species most commonly used for modern commercial maple syrup production because its sap has the highest concentration of sugars (2-3%). However, *A. rubrum* is also used commercially. The sugars in the sap of *A. negundo* are less concentrated.

Season – The timing of sap gathering is crucial, and changes slightly from year to year, depending on the temperature cycles. For sugar maple, it is usually in January for southern areas of the Northeast and as late as March or April in the northern range. The sap must be gathered when it starts to flow from the root stores into the aboveground parts. When the nighttime temperature is below freezing and the daytime temperature is above freezing, the sap will flow. Weather changes can have strong effects on the flow, consistency, and flavor of maple sap, and after budding, the sap becomes bitter.

The inner bark of boxelder may have a longer season of use, though it is likely of best quality in the same season as its sap running, with the sugars being most concentrated at this time.

Practice – The timing of sap flow of *A. negundo* in Central Texas is difficult to determine, especially as below freezing temperatures are rare in general in the region. I have attempted to tap boxelder for sap for several years in January, February, and March, all with no success, but there may also be errors in my technique in general. The southernmost tribe that tapped boxelder sap was the Kiowa in Oklahoma.

I eat the tender young samaras (winged seed pods) of boxelder, which are abundant and palatable. The leaves and other plant parts are non-toxic. I use the leaves to line earth ovens as they are relatively large and easy to gather in abundance.

Acer negundo. Base photo: full-size tree by a creek in early December. Overlay, clockwise from top left: soft, matte leaves with three (sometimes five) leaflets; flowers in late March; mature, two-winged samaras in early October; opposite branching and green bark; range map (GBIF).

ADOXACEAE – Muskroot Family

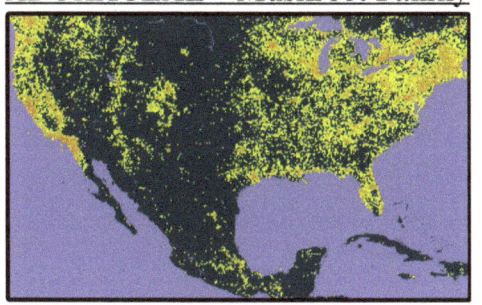

Sambucus distribution. Source: GBIF.

Sambucus spp.
Elderberry

Loc.: all TX except NW; common in Travis Co. (1 sp.); 2 spp. in TX (*S. canadensis* & *S. cerulea*). Form: shrub, tree; perennial.

Notes – Many sources, including the current USDA Plants Database and BONAP, have conflicts in the classification and naming of *Sambucus canadensis*, *S. cerulea*, *S. nigra*, and *S. racemosa*. *S. nigra* is native to Europe and has been introduced to parts of the northeastern US. *S. racemosa* is native to both the US and Europe, but does not occur in the wild in Texas, being mainly found in the Rocky Mountains and westward, northern Appalachians, and Great Lakes. *S. canadensis* is native to most of North America but does not occur in the Great Basin or Pacific Northwest (UT, NV, ID, OR, and WA). *S. cerulea* occurs only in west Texas (Big Bend area), and all states further west, especially in the Great Basin and Pacific Northwest. *S. mexicana* is also found in the US Southwest and Northern Mexico, but not in Texas. These specific epithets are often given as varieties or subspecies of one another (e.g., S. nigra ssp. cerulea = *S. cerulea*).

History

Fruits – Ripe elderberries were eaten by the Chiricahua and Mescalero Apache, Cahuilla, Calpella, Cherokee, Concow, Dakota, Flathead, Haudenosaunee, Houma, Kawaiisu, Kiowa, Klamath, Little Lake, Meskwaki, Omaha, Pawnee, Kashaya Pomo, Ponca, Shasta, Wailaki, Yokia, Yuki, Yurok, Natives in the Pacific Northwest and folk in northern Mexico. Species eaten include *S. canadensis*, *S. cerulea*, *S. racemosa*, and *S. mexicana*.

Elderberry fruits were eaten fresh and raw or were sun-dried for winter storage. They were made into pies, jellies, or jam (Cherokee, Flathead, Meskwaki). They were also eaten boiled or cooked into sauce (Flathead, Haudenosaunee).

The Cahuilla gathered large quantities. The fruits were dried for storage, then cooked into a rich, sweet sauce. They were stored in pottery jars for use year-round.

Elderberries were gathered in July and August (Cahuilla) or late summer (Kashaya Pomo).

Flowers – The flower umbels were dipped into hot water for a pleasant tea (Dakota, Omaha, Ponca, Pawnee). Flowers were gathered in early to mid-summer (Kashaya Pomo).

Notes

Character – *S. canadensis* is the only common elderberry species in Texas, except for *S. cerulea*, which occurs in limited areas of West Texas. It is the most widespread elderberry in the eastern half of North America. The five species of elderberries in North America share similar appearance, functional properties, and have historical Indigenous uses for food, medicine, and material. In Central Texas, elderberries grow wild in habitats with higher soil moisture, especially alongside creeks. They tend to grow in patches rather than individually.

Season – They flower from late spring to early summer and fruit from mid- to late summer.

Nutrition – The fruits of *Sambucus* species are rich in anthocyanins and much better in total antioxidant capacity than blueberry, mulberry, raspberry, and strawberry.[55]

Practice – I gather elderberry fruits by cutting the main stems of entire fruiting clusters, allowing the whole umbels to fall into a container. I gather the flowers in the same way.

      I then dry the whole umbels with the fruits attached. Once dry, I crumble them in my hands, causing the fruits to release from the stems. I put them on a flat tray, and winnow off the lighter stems by gently tapping or shaking the tray in a light breeze or with the aid of a fan.

      I store the dried fruits sealed in jars. They can be rehydrated by soaking in water, then they can be used as one would with fresh fruits. The flower umbels can be dried in a paper bag, and once dried, the flowers can be shaken off the stems inside the bag.

WARNING – The plant parts other than the ripe berries are poisonous. The plant contains cyanide compounds, including in its unripe fruits. Ripe fruits have lower concentrations of the cyanide, and they can be ensured to be safe to eat by cooking or drying them first. A study on *S. canadensis* found total cyanogenic potential to be lowest in ripe fruits compared to all other plant parts, and their fruits generally contain even less cyanogenic glycosides than those of *S. nigra*.[8]

Sambucus canadensis L.
American black elderberry

= Sambucus bipinnata, S. cerulea var. arizonica, S. eberhardtii, S. nigra ssp. canadensis, S. nigra var. canadensis, S. orbiculata, S. oreopola, S. plantierensis

American black elderberry, American elderberry, elderberry, elder, *sauco, sureau*

Arapaho: *kokúyono*, Cherokee: *koʔsagá / koʔs'agá*, Dakota: *chaputa,* Omaha-Ponca: *wagathahashka,* Pawnee: *skirariu,* Flathead: *ćkwik-alkshkw,* Gosiute: *pa'-go-no-gwĭp,* Haudenosaunee: *oniot'sŭtgŭs,* Hocąk: *hicocox* – "hollow stem," Menominee: *papaskitcî'ksi känax'tîk,* Meskwaki: *pakwana'mîshį[h] / papasikana'ʔtîk / popoki'mînûnį[h]* – "berry" / *pakwananoke' kotêk* – "berries of the elder," Osage: *bapoki hi* – "popping blackhaw plant"
Loc.: all TX; common in Travis Co.
Form: shrub, up to 12 ft. tall; perennial.
Flowers: April-July (white).
Food: Cherokee, Dakota, Flathead, Haudenosaunee, Houma, Kiowa, Meskwaki, Omaha, Pawnee, Ponca, northern Mexican folk.

Sambucus canadensis. Base photo: fruiting plant in mid-July (note opposite, pinnately compound leaves). Overlay, clockwise from top left: flower umbel; harvested fruit umbels in early August; stems (note lenticels/bumps and opposite leaf petiole scars); range map (GBIF).

ANACARDIACEAE – Sumac Family

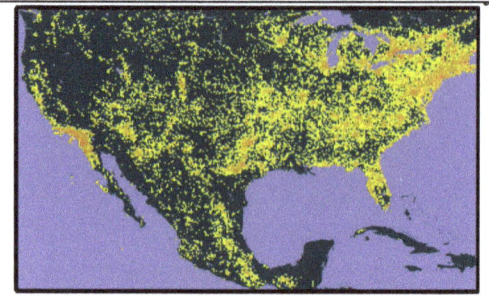

Rhus distribution. Source: GBIF.

Rhus spp.
Sumac

Choctaw: *bushucha*
Loc.: all TX except far S; very common in Travis Co. (4 spp.); 7 spp. in TX.
Form: low shrub, shrub, tree; perennial.

History

Fruits – The fruits of various sumac species were eaten by the Apache, Cahuilla, Comanche, Diné, Gosiute, Haudenosaunee, Hocąk, Hopi, Isleta, Jemez, Mexican Kickapoo, Kiowa, Menominee, Meskwaki, Ojibwe, Osage, Forest Potawatomi, Shasta, Tewa, and Mexican folk. Species eaten include *Rhus aromatica*, *R. glabra*, *R. integrifolia*, *R. microphylla*, *R. ovata*, *R. trilobata*, *R. typhina*, and *R. virens*.

Sumac fruits were commonly soaked in water to make a beverage (Western Apache, Cahuilla, Diné, Haudenosaunee, Hocąk, Hopi, Mexican Kickapoo, Kiowa, Menominee, Ojibwe). This drink has been called "Indian lemonade." The fruits were soaked whole, mashed, or dried and ground to make this drink. A sweetener, such as maple sugar, was sometimes added (Haudenosaunee, Hocąk, Meskwaki, Ojibwe). An infusion or decoction of the fruits was sometimes drunk (Kiowa, Menominee, Ojibwe).

The fruits were also sometimes simply eaten whole (Apache, Cahuilla, Kiowa, Meskwaki, Tewa). The unripe green fruits were eaten with salt (Kiowa). Mush or gruel made from the fruit meal was sometimes eaten (Cahuilla, Diné). The fruits were sun-dried, ground between flat stones, and the meal was boiled to make a jam that was often eaten with sunflower seed cakes or agave hearts (Chiricahua and Mescalero Apache).

The fruits or ground meal were dried for storage (Apache, Cahuilla, Diné, Hocąk, Kiowa, Menominee, Ojibwe, Forest Potawatomi, Tewa). Sumac fruits were an important part of the Kiowa diet. The sumac fruit gathering season was one of the six time periods the Chiricahua Apache used to divide the year (agave in the spring, then acorns, yucca fruit, sumac fruit, and other foods). The fruits were mostly gathered in the summer, but the fruiting season varies by species.

Notes

Character – Sumac fruits are tart, red, semi-fleshy, and have a fuzzy or pubescent texture. Poison sumac (*Toxicodendron vernix*) is in the poison ivy genus, and thus is not a true sumac (*Rhus*). Its fruits are smooth, whitish to greenish, hanging, and smaller than those of sumac. All true sumacs have edible fruits. Four *Rhus* species occur in Travis County, and all are common.

Season – *R. aromatica* and *R. trilobata* fruits ripen from early to mid-spring, *R. lanceolata* fruits ripen from late summer to early fall, and *R. virens* fruits ripen from mid- to late winter.

Nutrition – *R. aromatica* fruits contain 9% protein, 31% fat, 29% carbohydrate, 19% fiber, and 9% moisture.[60] They are about 40% flesh, with the remainder being an edible seed.[60] The fruits are much richer in polyphenols and carotenoids (46 mg/100 g fruit) than common foods such as blueberries and carrots.[60] They also contain malic acid (1360-2800 mg/kg), citric acid (50-95 mg/kg), and vitamin C (4-21 mg/kg).[77] It is this very high malic acid content, which is mainly coating on the exterior of the fruits, that gives sumac its characteristic sour flavor. Despite popular belief, they are only low to modest sources of vitamin C.

Practice – Of the four sumac species in Travis County, it is easiest to harvest the most fruits from *R. lanceolata*, as its fruits are borne in large, dense clusters, usually at the apex of branches. Fruiting clusters can simply be broken off the plant and the fruits later separated from the stems, which is easiest after drying. *R. trilobata* and *R. aromatica* fruits are borne in small clusters dispersed throughout the plant. They are easiest to harvest with the "strike and catch" method, but can also be easily hand-gathered. *R. virens* fruits are the largest, are borne in medium-sized clusters, and can be harvested by either method.

To make a drink with the fresh fruits, I lightly mash them in a mortar and pestle, mix the mash with water, and immediately strain it to make a delicious, sweet, and tart drink that I call "sumacade." For an even quicker method, I sometimes put whole fruit clusters (especially from *R. lanceolata*) in water, mix or lightly heat it, and then strain. For long-term use, I dry the fruits, grind them into a meal, then store them in a sealed container. This powder can be added to water with or without straining to make a sweet and tart drink. This powder can also be used as a cooking spice, as is done with sumac (*R. coraria*) in many Middle Eastern recipes.

Rhus aromatica Aiton & *Rhus trilobata* Aiton
Fragrant sumac & Skunkbush sumac
= Schmaltzia serotina, S. trilobata
Three-lobed sumac, lemonade berry, basketweed, *lemita, limilla*
Apache: *tciłtci* – "smelly wood," Cahuilla: *selet*, Northern Cheyenne: *hoʔ atoonoʔ ėstse* – "smoke issues," Comanche: *datsip*ʸ, Diné: *kʼįiʼ / tchiiłtchin* (*tchii* – "red"), Gosiute: *aiʼ-tcĭb*, Hopi: *cübi / sübi / su·ʼvi* (the plant) / *suviʼfsi* (the fruits), Jemez: *jı́npooh*, Kawaiisu: *iitʃivɨ*, Kiowa: *dtie-ai-pa-yee-ʔgo* – "bitter red berry," Tewa: *kun*, Zuñi: *koʼse oʼtsi* – "biting, man"
Loc.: all TX except far S; common in Travis Co.
Form: shrub, up to 12 ft. tall; perennial.
Flowers: Mar-June (white, yellow).
Notes – *Rhus trilobata* and *R. aromatica* are both accepted names, although they have been considered synonymous by some authors. Roughly, *R. aromatica* covers the eastern half of the US and *R. trilobata* covers the western half. They are very similar, and are confused in various sources, so I have combined them. Both occur in Travis Co., and can be distinguished by the glossier, smaller leaves of *R. trilobata* compared to the larger, more matte, and softer leaves of *R. aromatica*, though this trait occurs on a continuum and is partially environmental. These species are each other's closest genetic relatives[41,76] and for all purposes here, can be considered one.
Food: Apache, Cahuilla, Diné, Gosiute, Hopi, Isleta, Jemez, Kiowa, Shasta, Tewa.

Rhus lanceolata (A.Gray) Britton
Flameleaf sumac
Prairie sumac
Loc.: C, W, NE, & E TX; common in Travis Co.
Form: shrub, tree, up to 20-30 ft. tall; perennial.
Flowers: July-Aug (white, yellow, green).
Notes – *R. copallinum* and *R. lanceolata* are very similar. The former has glossy (shining) leaves, whereas the latter has matte leaves. Both have a winged rachis, unlike *Pistacia chinensis*.
Food: no records.

Rhus virens Lindh. ex A. Gray
Evergreen sumac
= Rhus choriophylla, R. sempervirens, Schmaltzia choriophylla
Tobacco sumac, *lambrisco, lentrisco*
Comanche: *temaichia*, Tonkawa: *ayumé*
Loc.: C, W, SW, & S TX; very common in Travis Co.
Form: shrub, tree, up to 6-10 ft. tall; perennial.
Flowers: June-Nov (white).
Food: Mexican folk.

Additional notes:
Rhus virens leaves were dried and smoked, usually mixed with tobacco, by the Comanche and Tonkawa. They were gathered in the fall, when the leaves turned red. The leaves of other sumac species, including *R. glabra*, *R. trilobata*, and *R. typhina*, were used similarly for smoking (Jicarilla Apache, Northern Cheyenne, Choctaw, Comanche, Coshetta, Dakota, Gosiute, Kiowa, Lakota, Omaha, Osage, Pawnee, Ponca, Forest Potawatomi, Tawehash, Tewa, Tonkawa, Waco, Wichita, Winnebago).

Top half: *Rhus virens*. Base photo: ripening fruits in early January. Overlay: flower in mid-October; fully-ripe fruits in early January; range map (GBIF).
Bottom half: *Rhus trilobata*. Base photo: ripe fruits and foliage in late May. Overlay: *R. aromatica* foliage (note it is larger, thinner, and more matte); combined range map of *R. aromatica* and *R. trilobata* (GBIF).

Rhus lanceolata. Base photo: fruiting plant in mid-August (note winged rachis of the leaves and lanceolate leaflets). Overlay, clockwise from top left: red foliage in early December (giving it the name "flameleaf sumac"); range map (GBIF); close-up of ripe fruits; dried harvested fruit clusters.

AQUIFOLIACEAE – Holly Family

Ilex vomitoria Aiton
Yaupon

Yaupon holly, cassina
Loc.: SE, E, & C TX; very common in Travis Co.
Form: shrub, tree, up to 25-45 ft. tall; perennial.
Flowers: Mar-May (white).
Notes – *Ilex vomitoria* and *I. cassine* have a history of being classified as the same species in scientific literature.[24] Cassine is common in Florida and its nearby coasts, whereas yaupon has a wider range in the southeast. Cassine has leaves that are about twice the size of those of yaupon.

History

Leaves – A decoction of yaupon leaves was drunk by the Asinai, Caddo, Cenis, Comanche, Creek, Karankawa, Tawokani, Tonkawa, Texas Natives in general, and other Southeast Natives. The preparation drunk was known as "Black Drink." Texians (early Texas folk) called the tree "*té del Indio*," "Indian tea," "chocolate," and "*chocolate del Indio*."

The "tea" was drunk as a decoction (the leaves were boiled) rather than as an infusion. Most accounts of the above Natives note that the foaming of the decoction was an essential component of the preparation, with the liquid was whisked or manually frothed to produce a yellowish foam on top. This foam was sometimes skimmed off, placed in a separate vessel, and drunk.

In Cabeza de Vaca's account from 1530 AD, Texas Natives regularly drank yaupon tea. These peoples (possibly Karankawa) first toasted the leaves in a pottery bottle. Once toasted, the bottle was filled with water and held over the fire "long enough to boil three times." The decoction was poured into a half-gourd bowl. When there was a lot of foam, it was drunk as hot and fresh as possible. The vessel in which the leaves were boiled was always kept covered. It could be drunk for three days without food, with each individual consuming about four and a half gallons per day.

Texas Natives (Asinai, Caddo, Comanche, Karankawa, Tawokani, Tonkawa) boiled the leaves in water and used "a bundle of small pieces of wood to make the decoction foam like chocolate." Referring to the foam, "this extract is placed in another vessel or reduced a second time to a liquid state." Around 1680, the Cenis made a decoction of the leaves that was "churned like chocolate, so that it also makes much froth," and drank it hot, "especially after they have walked a great distance."

Yaupon tea was drunk daily by Texas Natives, as well as in ceremonies wherein much larger amounts were consumed. It was drunk among the Asinai at the first fruits ceremony, at feasts, and by medicine men before doctoring. For example, the Asinai consumed Black Drink during a September new moon ceremony.

The Karankawa held a Black Drink ceremony at every full moon and after a very successful hunting or fishing expedition. They similarly "boiled a very strong and black decoction" that was occasionally "stirred with a sort of whisk, till the top was covered thickly with a yellowish froth." The Black Drink was freely drunk from an earthenware vessel that was passed around a circle. There was chanting in chorus accompanied by gourd rattles and flutes and the ceremony continued throughout the night.

It was probably used by every tribe within the plant's range, and even beyond. Inhabitants of Cahokia, a proto-urban center of the Mississippi, were drinking Black Drink around 1050 AD, as evidenced by beaker pottery residue.[18] Yaupon leaves were being imported from over 300 miles away, possibly from the Caddo, since the tree did not occur any closer and does not appear to have been cultivated at Cahokia.[18]

Notes

Character – Despite the specific epithet "*vomitoria*," Black Drink or yaupon tea does not have emetic properties.

It was named this by William Aiton in 1789 because it was associated with purgative ceremonies of the Natives, such as was depicted in famous engravings of the Timucua in Florida (De Bry 1591). In such ceremonies of Southeast Natives, one intent was to cleanse the body by voluntarily purging, whether by mechanical stimulation, addition of actual emetic compounds (in some versions of Black Drink), or simply the result of consuming heroic volumes of the drink. This association of the plant with vomiting was the only reason Aiton had for giving the plant this name. Living in Kew, UK, having only seen specimens of the plant and never tried Black Drink, he perhaps believed it caused vomiting based on limited information such as the engravings and underinformed accounts. Several myths are commonly perpetuated surrounding the naming of this plant, with no substantiation. It was named along with hundreds of other species by Aiton and others in the *Hortus Kewensis*, volume one.[3] In this volume, the common names given for *I. vomitoria* were "South-Sea Tea, or Ever-green Cassine."[3] The type specimen was from Florida.[3]

Possumhaw (*Ilex decidua*) often occurs intermixed in the same habitat as yaupon and looks very similar. Possomhaw leaves have a tapered base and are deciduous. Yaupon leaves are ovate (completely oval-shaped, lacking a tapered base) and evergreen.

Nutrition – Yaupon leaves have about 40% of the caffeine of coffee beans and about 10% of the theobromine of cocoa beans.

Yaupon leaves contain a median 0.56% caffeine by dry weight, compared to 3.4% in common tea (*Camellia sinensis*) leaves and 1.4% in coffee (*Coffea arabica*) beans.[24] Cassine (*Ilex cassine*) leaves contain 0.12% caffeine.[24] Yaupon leaves contain five times the caffeine of cassine leaves.[24]

Yaupon leaves have 0.11% theobromine,[24] compared to about 1% in cocoa (*Theobroma cacao*) beans. Cassine leaves contain 0.22% theobromine: twice the amount as yaupon leaves.[24]

Black Drink contains caffeine and theobromine in a ratio of about 5:1.[18] In cocoa, the ratio is approximately reversed.[18]

Season – Yaupon is evergreen, and can be gathered year-round. It is very common in Austin, most often growing in moist soils, especially alongside creeks.

Practice – I have prepared yaupon tea many hundreds of times, often making it as strong as possible and drinking as much as I can on an empty stomach, and it does not produce any emetic sensations. Any experienced yaupon drinker can corroborate this.

I gather yaupon by stripping branches of leaves with my hands. It is easier to strip long, straight branches. I do not discriminate between young and older leaves, but it is likely that the mature leaves contain more caffeine. I either sun-dry the leaves or roast them on a baking tray at a low oven setting until they are crispy and slightly brown. Air- or sun-dried green leaves are perfectly usable for tea, but the fresh leaves are difficult to extract. I sometimes smoke or roast them on a fire, which adds to the flavor. I crush up the dried leaves with my hands.

I put about a cup of dried, crushed leaves into two quarts of boiling water and let it boil vigorously for about ten minutes. I then strain out the leaves and drink it hot, usually about a pint at a time, but sometimes as much as two quarts. Its effects are comparable to yerba maté (*Ilex paraguariensis*), guayasa (*I. guayusa*), or a combination of coffee and chocolate.

A layer of foam is usually naturally produced at the top, but whisking can increase the foam. I sometimes skim off the foam, place it into another vessel, let it settle, and drink that. This is like the espresso of yaupon, being more potent than the liquid fraction of the decoction.

Ilex vomitoria. Base photo: fruiting plant in early December. Overlay, clockwise from top left: Black Drink being poured; flowers in late March; range map (GBIF); foliage (note that the leaves are simple, ovate, thick, glossy, and have wavy margins).

BERBERIDACEAE – Barberry Family

Berberis trifoliolata Moric.
Agarita

= Alloberberis trifoliolata, Berberis trifoliata, Mahonia trifoliolata
Algerita, agrito, wild currant, currant-of-Texas, mahonia
Loc.: all TX except far E & far N; common in Travis Co.
Form: shrub, up to 10 ft. tall; perennial.
Flowers: Feb-Apr (yellow).

History

Fruits – Agarita berries were eaten by the Mexican Kickapoo, especially made into jelly. The berries of other *Berberis* species were eaten by the Apache, Blackfeet, Northern Cheyenne, Flathead, Jemez, and Natives in the Pacific Northwest.

Notes

Character – These are some of the best wild fruits in Texas and are often produced prolifically. This shrub is common and grows in many habitats but prefers forest edges and other semi-open areas. It is very easy to identify with confidence. The thick, spiny, glossy, trifoliate leaves (with three leaflets) are not similar to anything else, except perhaps American holly (*Ilex opaca*), which usually occurs only as an ornamental where agarita grows, and has simple leaves rather than three leaflets.

The berries are sweet, tart, juicy, and fruity fresh and raw. Their tiny seeds can be eaten whole along with the fruits, which can be easily prepared into jelly, jam, preserves, juice, or even wine.

Season – Agarita fruits ripen in early to late spring, and sometimes last into early summer, but are mostly ripe in May.

Nutrition – The fresh fruits of *Berberis heteropoda* contained 3% protein, 1% fat, 18% carbohydrate, and 87 kcal/100 g.[62] They also had high antioxidant activity and contained potassium (583 mg/100 g), calcium (79 mg/100 g), phosphorus (73 mg/100 g), and magnesium (31 mg/100 g).[62] They are an excellent source of potassium and a moderate source of phosphorus, calcium, and magnesium.

Practice – These fruits are difficult to gather by hand because they are small and the leaves have stout prickles. With the "strike and catch" method, and a good fruiting year, I can gather about 20 pounds of agarita fruits in an hour. I devised the foraging net described in the "Gathering" chapter specifically to gather agarita fruits, though it works for many others. I also refined the "water cleaning" method for agarita fruits. I eat them fresh, can them as jelly, or dry them for storage. I put the dried fruits in smoothies or rehydrate them by soaking for various uses.

Berberis trifoliolata. Base photo: fruiting plant in late April. Overlay, clockwise from top left: ripe fruits in early May; range map (GBIF); yellow wood (which is an excellent dye); harvested and cleaned ripe fruits in mid-May, flower in early March; glossy, thick leaf with three spiny leaflets.

BORAGINACEAE – Borage Family

Ehretia anacua (Terán & Berl.) I.M. Johnst.
Anacua

= Ehretia eliptica, Gaza anacua
Sugarberry, sandpaper-tree, knockaway, *anaqua, manzanita, manzanillo, tlalahuacate*
Loc.: S, SE, & C TX; common (often ornamental) in Travis Co; only *Ehretia* spp. in TX.
Form: tree, up to 50 ft. tall; perennial.
Flowers: Feb-Apr (white).

History

Fruits – These were eaten by Texas Natives and the Mexican Kickapoo.

Notes

Character – The small, sweet, juicy, and yellow to red fruits of anacua are produced in incredible abundance on densely-clustered infructescences covering these large trees. The fruits have a pair of medium-sized seeds which are quite hard but can be cracked in the teeth and eaten, especially when dried. The tree is somewhat oak-like in appearance, with simple, ovate, alternate, deciduous leaves. The leaves have an extremely rough texture, giving it the name "sandpaper-tree."

Season – The fruits are available from May to August, with peak ripeness in late June to early July.

Nutrition – The flesh of *Ehretia tinifolia* fruits from Mexico contain 17% protein, 0% fat, 5% fiber, 293 kcal/100 g, selenium (12 µg/100 g), and magnesium (10 mg/100 g).[50] The high protein content of the pulp is remarkable, and the fruits are a good source of selenium. *E. anacua* fruits appear to be very similar and may have a comparable nutritional composition.

Practice – I gather these with the "strike and catch" method, using my foraging net (see "Gathering" chapter). The fruits are typically not all ripe at once, so I harvest from a tree, wait a few days, then harvest from it again, repeating as desired while the fruits remain. A lot of debris is often gathered with this method, especially during the first harvest.

I pull out the large sticks and leaves by hand, then spread the fruits in an open spot outdoors to let the bugs walk off. Next, I winnow off the light debris. Then, I use the "water cleaning" method to remove everything except the fruits. After draining the water, they are ready to be eaten fresh, dried, or processed further.

To remove the seeds, which I prefer to not eat, I put the fruits in a pot with enough water to cover them. I cook this on low, occasionally stirring and agitating the fruits, until they are reduced to a pulp. I strain the pulp through a wire mesh sieve to separate the seeds, then can the pulp or drink the juice. I eat the dried fruits whole or rehydrate and prepare them as I would fresh fruits.

Grinding the whole dry fruits may also be an option, depending on the edibility of the seeds. I have eaten many of the seeds with no ill effect. The seeds of other genera (*Agastache*, *Amsinckia*, *Cordia*, *Lithospermum*, *Nama*) in the borage family were eaten by the Gosiute, Kawaiisu, and Mexican folk.

Ehretia anacua. Base photo: fruiting plant in late June. Overlay, clockwise from top right: range map (GBIF); rough bark; ripe, red fruit; simple leaf with a visible sandpaper texture; harvested and cleaned ripe fruits.

CANNABACEAE – Hemp Family

Celtis laevigata Willd.
Sugarberry

= Celtis mississippiensis, C. reticulata, C. smallii
Sugar hackberry, Texas sugarberry, southern hackberry, lowland hackberry, *palo blanco*
Apache: *iyntɬidz* – "hard seed," Comanche: *natsɔkwɜ / miːtsɔná / mitsonaaʔ*, Kawaiisu: *nawakkimahavi*, Tohono O'odham: *kom*, Tewa: *p'ekeʔin* – "hard stick"
Loc.: all TX; very common in Travis Co.
Form: shrub, tree, up to 80 ft. tall; perennial.
Flowers: Feb-Apr (green).

History

Fruits – Sugarberries were eaten by the Apache, Comanche, Tewa, and Tohono O'odham. They were gathered in large quantities (Tohono O'odham). The fruits from other *Celtis* species (*C. ehrenbergiana* and *C. occidentalis*) were eaten by the Dakota, Diné, Mexican Kickapoo, Kiowa, Meskwaki, Omaha, Osage, Pawnee, Seri, Texas Natives in the Archaic, and folk in northern Mexico.

They were sometimes eaten whole and raw (Apache, Seri) but were usually ground into a meal or paste (Apache, Comanche, Dakota, Diné, Kiowa, Meskwaki, Osage, Pawnee). The ground fruits were formed into cakes and dried for storage (Apache). The paste was mixed with fat (and sometimes with cornmeal or sugar), molded onto the end of sticks, and roasted over a fire (Comanche, Kiowa, Pawnee). The meal was also eaten as a mush (Meskwaki), jelly (Apache), or meat flavoring (Dakota).

Hackberry (*Celtis* spp.) fruit remains have been found at numerous archaeological sites in the Southeast and Texas, spanning from the late Paleoindian and Archaic periods to historical times, and made up a substantial part of the diet for many Native peoples.[15]

Leaves – A decoction of *C. occidentalis* leaves was drunk as a "spring tonic" (Kiowa).

Shoots – The Caddo used *Celtis* shoots for basketry used in food processing as the wood does not impart any taste.

Notes – The Dakota name given for *C. occidentalis* was "*yamnumnugapi*," with "*yamnumnuga*" meaning "to crunch," in reference to the hard crunch of the seed pits when one chews hackberry fruits.

Notes

Character – These are some of the most common trees in Austin and other urban areas, especially on fencelines, as the fruits are favored by birds that perch on fences and deposit the seeds, which grow there protected from mowers. They are also very common in the wild.

They are very easy to identify from just their bark. The bark has raised protuberances or ridges of stacked, corky tissue, with smooth spans of interstitial bark. The specific epithet "*laevigata*" refers to this smooth texture, distinguishing the species from *C. occidentalis* (common hackberry), which has less of these smooth spans of bark.

The fruits consist of a thin layer of sweet, orange-red flesh covering a single, very hard endocarp (pit), which encloses a soft seed.

Season – The fruits ripen from mid-summer through fall, often remaining on the tree until late winter.

Nutrition – Sugarberry endocarp contains mostly calcium carbonate crystals, 70-99% of which are aragonite.[15, 59] Whole sugarberries contain 11% protein, 10% calcium, and 72 mg/100 g magnesium.[26] They are a good source of magnesium, providing the same amount as cashews. Whole *C. australis* fruits contain 7% fat.[20] The flesh alone contains 50% carbohydrate.[47] The nutrient composition of the flesh is similar to dried figs, but has five times the carotene concentration.[47] The flesh is rich in iron, with 5 mg/100 g,[47] more than flour. The fruits are also a good source of Mn, Zn, Na, K, P, and Ca.[20]

Practice – The young flowers, buds, and young leaves are edible raw. In the spring, just after leafing out, these parts are available and have a mild herbal taste. The larger leaves, when still soft and young, furnish a pleasant tea. All parts of the fruit are edible raw and they can be dried whole for long-term storage.

I gather large amounts of the fruits with the "strike and catch" method. These trees are very common in urban areas, and I have gathered many from upper branches by standing on a roof. Otherwise, I am restricted to what I can reach. I winnow off the debris, which is sufficient to fully isolate and clean the fruits because they are so dense. Many fruits remain on the branches after the leaves have fallen and can be gathered at this time with less need for winnowing, although they are less fresh.

The fruits are good raw, but their seeds are unpleasantly hard to chew. I still chew and eat them whole, but prefer not to eat many this way. I sometimes chew off the sweet, fleshy part and spit out the seeds. When I gather a large quantity, I grind them into a paste with a mortar and pestle. A food processor can speed this grinding. I mix this paste with coconut oil and honey, form little cakes and bake them in the oven like cookies. They are delicious.

I also use hackberry shoots and wood for cooking utensils and other implements.

Celtis laevigata. Base photo: foliage in early summer. Overlay, clockwise from top left: warts or ridges of stacked, corky bark with smooth spans in between; range map (GBIF); harvested ripe fruits in early November; ripe fruits on tree in late October.

EBENACEAE – Ebony Family

Diospyros texana Scheele
Texas persimmon

= Diospyros cuneifolia, D. mexicana, Brayodendron texanum
Mexican persimmon, black persimmon, *chapote, sapote, chapote prieto, sapote prieto, nísperos*
Comanche: *dunaseika / tuhnaséka*
Loc.: C, SW, S, & SE TX; common in Travis Co.
Form: shrub, tree, up to 45 ft. tall; perennial.
Flowers: Feb-Apr (white, green).
Notes – The Comanche name is equivalent to "*tuh-Diospyros virginiana*," and "*tuhupitʉ*" is the color black, with the prefix "*tuh*" being used to denote black, as in "*tuhmubitai*" – "black walnut." Thus, *tuhnaséka* means "black persimmon."

History

Fruits – Texas persimmons were eaten by the Apache, Aranama, Comanche, Karankawa, Mexican Kickapoo, Tonkawa, and folk in northern Mexico. They were highly esteemed by Natives in Central Texas and commonly eaten by Texas Natives in the Archaic.

They were most likely simply eaten fresh. They may have been preserved using methods similar to those described for *Diospyros virginiana* below, with the seeds removed and the pulp dried by the fire or sun.

Celiz and Hoffmann (1935) refer to these as "*medlar*," but *D. virginiana* does not occur in the territory of the Aranama. *D. texana* does, and the timing (late September) only matches the latter species.

Notes

Character – These are the best-tasting fruits I have ever eaten, with a rich, sweet, and deep flavor like molasses and prunes, a pudding-like texture, and a dark purple color. Their closest relative is *Diospyros nigra*, called "chocolate pudding fruit" or "*sapote negro*," which has similar (but larger) fruits with dark flesh that were cultivated by Indigenous Mexicans in the Tehuacán valley, Puebla, Mexico between 7,000 to 500 years ago.

The trees are most easily recognized by the bark on their trunks and large branches. The bark is smooth and light gray, with rougher, darker patches that peel away but often remain attached in a curved form. This smooth, light bark is unique in Texas and is only comparable to the introduced crepe myrtle, which is light brown, or the Texas madrone, which is light orange-brown.

The simple, alternate, ovate leaves look much like live oak but are more pubescent (fuzzy). The leaves can also be distinguished from live oak by the presence of distinct small, green, fuzzy galls (spheres) often protruding from the leaf tissue.

The heavily skewed male-to-female ratio makes fruiting trees about nine times less abundant than males. Female trees can be recognized by their fruits or bell-shaped, light yellow to white flowers. Male flowers are inconspicuous.

Season – The fruits start ripening in midsummer, with a peak ripeness from late summer (mid-August) to early fall (mid-September). They become difficult to find in October and are completely gone by November. Prickly pear fruits are also available in this season and in the same habitat, so I often gather them at the same time. Wherever they can be found abundantly, these two fruits are prime foraging targets.

One easy way to determine the availability of Texas persimmons in the field is to look for dark scat. Coyotes, raccoons, and many other mammals are fond of the fruit, which turns their scat a recognizable dark color and contains smooth, large, brown Texas persimmon seeds. It will start showing up at the very beginning of the season, when most fruits are still green. At peak ripeness, this scat is common. Coyotes have no qualms about eating up dried and spoiled fallen fruits, so this scat can be found later in the season than most people would consider the fruits themselves palatable.

Overripe Texas persimmons tend not to go "bad" or develop fungal growth. These sweet, moist fruits somehow remain relatively good to eat even after hanging on the trees for months. I find that they never really rot or get moldy; they simply shrivel up and dry. Some do start to taste "off" (slightly fermented) a few weeks after ripening, but I still find them enjoyable. This longevity is likely due to their defensive chemicals, especially an antifungal protein that has been isolated from the overripe fruits.[74]

Nutrition – The pulp of Texas persimmons contains 4% protein, 148 mg/100 g potassium,[26] and plenty of sugars.

Practice – I find the fruits are best gathered by simply picking them by hand. They are large enough that no more elaborate method is necessary. The "strike and catch" method does work, but hand-picking avoids the need to separate out debris collected in that method.

They are incredibly good plain and raw, but can be processed for preservation. I fill a pot with the fruits and cover them with just enough water. I cook this on a very low simmer for hours as the fruits slowly fall apart, agitating them with a spoon or similar tool to aid disintegration. Once the batch becomes a slurry of pulp, seeds, and skins, I strain out the seeds and skins using a wire mesh strainer. The strained pulp retains flavor and sweetness through several rounds of this process, so I put it back in the pot with more water. Because I need to add water to keep extracting the pulp after each round of straining and shaking, the separated pulp is typically more watery than I want. So, I cook it in the oven on low (200° F or lower) or over a low flame on the stovetop to evaporate the excess water, until it reaches the consistency of a thick pudding. I then can it hot or use it in recipes. My favorite recipe is 50:50 Texas persimmon pulp and coconut cream, blended together and then put in an ice-cream maker. It needs no sweetener and has the perfect texture, color, and taste.

Diospyros virginiana L.
Common persimmon

American persimmon, date plum, possumwood, *medlar, piacminiar, nisperos*
Cherokee: *sulí / salí / salí,* Comanche: *nase'ka / naséka,* Osage: *θta-iŋge,* Quapaw: *piaquiminia*
Loc.: E & sparse C TX; not uncommonly cultivated, but rare in the wild in Travis Co.
Form: tree, up to 100 ft. tall; perennial.
Flowers: Mar-June (yellow, green).

History

Fruit – Common persimmons were eaten by the Aranama, Asinai, Caddo, Choctaw, Comanche, Karankawa, Nabedache, Natchitoches, Nazones, Osage, Tonkawa, Texas Natives in 1709, Natives in Arkansas, Illinois, and Oklahoma, and likely others.

The ripe fruits were beaten to a pulp, the seeds were removed, and the paste was spread out to dry (Comanche). The dried cakes were later softened in water and eaten in various preparations. The fruits were also "squeezed out" to dry and preserve them (Comanche).

The Osage covered a board measuring about one square foot with bison grease, upon which they placed three or four layers of whole persimmons. The board was held over a fire until the fruit "cooked" or was dried into a cake. After cooling, they were stored in rawhide pack containers. Each family prepared several such packs each fall and used them throughout the year.

The Asinai used dried persimmon paste or dried pulp cakes as trade items. The fruits were made into a bread by the Choctaw or added to venison stew. They were cooked into preserves by Natives in Oklahoma and Arkansas.

The fruits were eaten by members of the La Salle expedition to Texas. The Aranama, Tonkawa, and Natives in south-central Texas may have instead been eating *Diospyros texana.* Species ranges suggest *D.*

texana was more likely the species for the Aranama. The gathering date also suggests *D. texana.* Both species are found in Tonkawa territory. The territory of Natives in south-central Texas is outside the current range of *D. virginiana*, but *D. texana* is common there. The quote "in some rivers *medlar* trees are found like those in Spain" (Tous and Foik 1930) suggests *D. virginiana*. It is possible *D. virginiana* had a further westward range historically, or that authors were may have neglected to distinguish between it and *D. texana*.

Notes

These fruits are extremely sweet, with a delicious, juicy, gelatinous flesh. However, they should only be eaten when fully ripe. At this stage, they will be soft and squishy, like a water balloon. If they have firmness beyond that, being even slightly unripe, the taste is extremely astringent and unpleasant. They ripen from fall through winter. In Austin, they're mostly found in landscaping, but do occur in the wild in certain areas. Their leaves furnish a palatable tea high in vitamin C.

Diospyros virginiana. Foliage and ripening fruit in mid-October. Range map: GBIF.

Diospyros texana. Base photo: fruiting plant in late August. Overlay, clockwise from top left: ripe fruit in late August; female flowers in late March; range map (GBIF); smooth, gray bark on large tree; peeling bark on small tree; harvested ripe fruits in late August.

ERICACEAE – Heath Family

Arbutus xalapensis K. Kunth
Texas madrone

= Arbutus texana
Madroño, amazaquitl, xoxocote
Loc.: C & W TX; common in W Travis Co.; only *Arbutus* sp. in TX.
Form: shrub, small tree, up to 30 ft. tall; perennial.
Flowers: Feb-Apr (white).

History

Fruits – The fruits, tasting sweet like strawberries, were eaten by the Mexican Kickapoo.

The fruits of *Arbutus menziesii* were eaten by the Karuk, Kashaya Pomo, Yurok, and other Natives in Northern California, but could cause vomiting if eaten in large quantities. They were not considered good for storage, as the fruits decay rapidly when bruised unless first parched (Kashaya Pomo). They were roasted by the Yurok over an open fire before eating. The Karuk steamed the fruits before drying them for storage and soaked them before eating.

Notes

Character – This is one of the most beautiful trees in the Edwards Plateau (Hill Country), with smooth, light orange-brown, peeling bark. It has large, leathery, tropical-looking leaves. The fruits are aggregates, like raspberries, and are sweet and fruity, similar in taste to those from *A. menziesii*.

In Central Texas, they are at the far northern end of their range, being more of a tropical plant found as far south as Costa Rica. I presume environmental stress from the relatively extreme habitat prevents them from commonly flushing with fruits, as can be seen in California madrones (*A. menziesii*). They may flourish in certain habitats and furnish enough fruits to harvest as an appreciable food source, but in general, we should allow birds to eat the fruits, as this arresting tree actually cannot be grown in cultivation.

Season – The fruits ripen in late summer through winter, especially in late October.

Practice – I have tasted them, but never found the fruits in enough abundance to warrant any gathering.

Arbutus xalapensis (in Uvalde Co.). Base photo: smooth, peeling, orange bark and foliage. Overlay, clockwise from top left: unripe fruits (ripen to orange) in early July; range map (GBIF).

FABACEAE – Legume Family

Cercis canadensis L.
Redbud

= Cercis mexicana (= C. canadensis ssp. mexicana)
Eastern redbud, western redbud
Cherokee: *kwaniyusti* – "like peaches," Kiowa: *kee-à-gu-la*, Osage: *žoŋššabeðe hi* – "dark-wood tree"
Loc.: SW, C, E, & sparse N TX; common in Travis Co.; only *Cercis* sp. in TX.
Form: shrub, tree, up to 30 ft. tall; perennial.
Flowers: Feb-May (pink).
Notes – This species has previously been considered a synonym of *C. orbiculata* and *C. occidentalis*, both of which are now currently accepted. *C. occidentalis* may be found cultivated.

History

Flowers – Redbud flowers were eaten by the Cherokee; children were especially fond of them. In San Luis Potosi, Mexico, Indigenous peoples fried the flowers as a rare delicacy. French Canadians used them in salads and pickles.

Fruits – The fruit pods of *C. orbiculata* were roasted in ashes and the seeds (beans) were eaten by the Diné.

Notes

Character – Redbuds are among the first plants to flower in the year, blooming before they bud leaves, producing a vibrant sight of bright pink trees amid the otherwise austere landscape. Their light magenta flowers are pea-like and often emerge from the trunks and branches (cauliflorous).

Season – The trees flower in late winter to early spring, from February to April and into May. The bean pods are produced from early to late summer, with the beans ripening toward the end of that.

Practice – Redbud flowers are lightly sweet and faintly astringent, with a good flavor fresh and raw. They also make a nice tea. I like to eat them straight from the tree or added to salads or desserts. To store them, I dry them, using the dried flowers as tea.

The beans (seeds) are small, but it is not difficult to get a lot of them. The bean pods are produced in abundance, and each one has a handful of tiny beans. I roast the whole pods on a tray in the oven until they are browned, dry, and crispy. I crush the dried pods between my hands to reduce them to debris and release the beans. I winnow off the debris, leaving the beans behind. They are best with a bit of salt, and have an acceptable flavor overall.

Cercis canadensis. Base photo: flowering branch in late February. Overlay, clockwise from top left: heart-shaped leaf; range map (GBIF); harvested flower clusters in mid-March; mature bean pods in late April.

Neltuma glandulosa (Torr.) Britton & Rose
Honey mesquite
= Algarobia glandulosa, Ceratonia chilensis, Prosopis chilensis, P. juliflora, P. glandulosa, P. odorata, Mimosa pseudo-echinus
Glandular mesquite, mesquite bush, algarroba, *mezquite, mizquiqui, mezquite blanco, mezquite amarillo, mezquite colorado, chachacha, tahi, algaroba*
Aztec / Nahuatl: *mizquitl*, Akwa'ala: *enal*, Apache: *nastane* – "that which lies about," Cochimí: *guatrá*, Cocopah: *anyaʟ*, Comanche: *namoβitsɔni / natsɔkwe*, Kamia: (the pods) *anaxi*, Kawaiisu: *opiṁbi*, Seri: *haas*, Shoshoni: *oh vea*, Tewa: *tsep'e* – "eagle plant," Yukaliwa: *ahaᵃ*, Yuma: (the pods) *eya"*
Loc.: all TX; very common in Travis Co.; 2 *Neltuma* spp. in TX (*N. velutina*).
Form: shrub, tree, up to 30 ft. tall; perennial.
Flowers: Feb-Sept (yellow).

History

Water – Mesquite groves served as an indicator of ground water to the Cahuilla, who often dug wells in such groves.

Fruits – Mesquite pods were an important food source of the Chiricahua, Lipan, and Mescalero Apache, Cáhita, Cahuilla, Coahuiltecans, Cocopah, Hopi, Kamia, Kiowa, Kumeyaay, Maricopa, Moqui, Mohave, Southern Paiute, Pima, Timbasha, Tohono O'odham, Walapai, Yaqui, Yavapai, Yuma, Plains-Rockies Natives, Natives of Southern California in the mid-1700s, 1775, and in 1852, Natives of Baja California in the late 1700s, Indigenous peoples of New Spain in 1790, Southwest Natives, Mexican folk, Indigenous Mexicans, Indigenous Mexicans in 1500, Texas Natives 5,000 years ago, Texas Natives in the Archaic, and many historical Natives of Texas.

They were also eaten by the Comanche, Kawaiisu, Mexican Kickapoo, Kiliwa, Shoshone, folk in Northern Mexico, and Texas Natives called "Avavares" around 1530.

Honey mesquite was historically the most important wild edible plant for Indigenous peoples in the Southwest. Prickly pears (*Opuntia* spp.) come close, but include many species. Acorns were the most important among Indigenous North Americans in general, but were less important than mesquite in the Southwest. Honey mesquite also had more uses in general than any other species in the Southwest.

It is essential to understand the pod anatomy (see p.107). The outer part of the pod, specifically the mesocarp, the spongy tissue surrounding the beans, was the main source of food, not the beans inside. The beans (seeds and endocarp/shell) were not commonly eaten by Native peoples historically.

A staple sweet beverage, which I call "mesquite drink," was made from the bean pods. It is very easy to prepare: simply smash and soak the pods. There were many methods, but typically, the bean pods were pounded in a large wooden mortar made from a softwood tree trunk using long hardwood pestles while standing. Bedrock mortars or mano and metate were also used for pounding mesquite pods. After pounding, the pulp was soaked in water, strained, and the cold infusion was drunk. This drink was sometimes allowed to steep in the pulp for days, and it became mildly fermented, improving the taste and possibly the digestibility.

This was only one of the two main ways mesquite pods were historically eaten. The other way was to make a meal or flour from the mesocarp. The pods were thoroughly dried, pounded, sifted, and winnowed to remove the beans, skin, and fibers from the meal. This meal was prepared in various ways, such as a mush, or formed into cakes that were dried for storage.

The beans were only sometimes eaten by Native peoples historically, and that required the removal of the endocarp (the outer "shell" of the bean). The beans were pounded, the shells winnowed off, and the seeds ground into a meal that was cooked. The effort to shell, grind, and cook the small seeds, compared to simply extracting the sweet mesocarp, was why the beans were usually discarded.

The bean pods were stored in large basketry granaries to deter rodent infestation. These were often made from aromatic plants or sealed with mud to deter insect infestation. There were many methods of gathering, processing, and preparing them, organized below by culture.

Apache – The bean pods were pounded into a pulp using a depression in a rock as a surface. The pulp was soaked in cold water, then was squeezed by hand or in a basket and strained out of the sweet drink. The Apache also made cakes of the dried meal.

An alternate method of preparation among the San Carlos Apache was to let the whole bean pods dry, discard the beans, and pound the pods (mesocarp) into pulp. This pulp was mixed with cold or warm water and eaten as mush, without cooking. Salt was sometimes added.

Chiricahua & Mescalero Apache – Pods were gathered, boiled, pounded on a hide or ground up on a metate, and the mixture was placed in a pan and worked with the hands until it reached a thick consistency. Sometimes, the pods were boiled until red, removed and mashed (and the beans likely removed), then returned to the container and cooked until most of the water had evaporated, creating a sort of pudding.

A fermented, mildly alcoholic drink was made by finely grinding cooked mesquite pods and adding a little water during the grinding process. More water was then added to the ground mass, and the mixture was allowed to ferment for a day and a night.

Lipan Apache – The Lipan Apache stored provisions of mesquite meal (likely just powdered or caked mesocarp tissue) in sacks for the winter. They called this food *piñole de mesquite*.

Cahuilla – It takes three or more weeks for full-sized green pods to become fully dry and mature. Green mesquite pods were gathered in early to mid-summer, and mature, naturally dried pods were gathered in early fall. Green pods were prepared for eating immediately after being picked or were sun-dried. Green or mature pods were pounded or crushed in mortars to produce a pulpy juice. In every Cahuilla house, a large clay basin containing half-crushed pods was kept filled with water, and the beverage was drunk regularly.

To pound mesquite pods, the Cahuilla used mesquite mortars and stone pestles in the same way as the Yuma. Mortars used were made from mesquite or cottonwood (*Populus deltoides*) stumps measuring two to three feet high. The mortar was prepared by burning out the center and then cutting a depression out with a stone axe. The typical mortar was about thirty inches high, with a bowl fifteen inches deep. A wooden or stone pestle, two to three feet long, was used while standing.

The Cahuilla placed mesquite meal in a basket, dampened it with water, and left it there for a day to harden. The meal was sometimes formed into round balls, but usually it was molded into cakes measuring two to ten inches diameter and one to three inches thick.

They tended mesquite groves by pruning, breaking, and cutting branches regularly to increase access to beans. Groves were owned by families, and the use of mesquite pods as a food source may have been the main factor allowing them to be sedentary.

It was estimated that, in a good season, a tree may produce an average of fifteen to thirty pounds of pods per year. A single worker could gather about 175 pounds of pods per day, and one acre of land well-covered with mesquite could yield about 3,000 pounds of pods annually. However, these estimates disregard that the Cahuilla were selective about which trees they harvested, as individual trees vary in pod palatability, ranging from very bitter to very sweet.

Whole dried pods were stored in basketry granaries for a year or possibly longer. Mesquite pod meal was kept in clay or basketry vessels, granaries, and dry caves. Dried cakes were stored by hanging them in rafters or in cloth bags. The Cahuilla in the Coachella Valley made granaries almost exclusively from white sagebrush (*Artemesia ludoviciana*). In other Cahuilla granaries, willow (*Salix* spp.) shoots, arrowweed (*Pluchea sericea*), or sagebrush were twisted into long ropes that were coiled upward, much like coiled basketry. The inside was plastered to make the granary airtight. After filling with mesquite pods, the granaries were sealed with sagebrush shoots and mud daub to prevent insect infestation. The granaries were set on platforms of poles or atop high boulders to keep them out of reach of small rodents. They were large enough to hold 300-500 pounds of beans, sufficient to feed a family of six to ten people for a year.

Mesquite was a regular trade item of the Cahuilla, and surplus pods or cakes were traded for acorns or other valued food resources from neighboring tribes such as the Serrano, Luiseño, and 'Iipai-Tiipai. Mesquite pod cakes were often exchanged with the Kawaiisu for pinyon pine seeds and acorns.

The Cahuilla named five out of eight seasons in their calendric system based upon the development stages of the mesquite pods: *Taspa* – "budding of trees," *Sevwa* – "blossoming of trees," *Heva-wiva* – "commencing to form pods," *Menukis-kwasva* – "ripening time of pods," and *Merukis-chaveva* – "falling of pods" (followed by *Talpa* – "midsummer," *Uche-wiva* – "cool days," and *Tamiva* – "cold days").

Cocopah – Care was taken to find the mesquite trees with the sweetest and fullest pods by breaking and tasting the pods. A long, hooked pole was used to pull down higher mesquite branches. This pole had a short crosspiece lashed with mesquite bark at an acute angle for a hook. Rat (probably the packrat, *Neotoma* spp.) nests were raided for mesquite pods and screwbeans. The Cocopah never used the beans or seeds of mesquite.

Dried pods were pounded in a mortar, with some of the beans and heavy fibers removed during grinding. The remainder was separated by shaking the ground meal in a basket to sift out the larger pieces. The resulting pod mesocarp meal was made into a drink by mixing it with water, or it was eaten dry and washed down with water. The meal could also be soaked in water, chewed to extract the juice, and the solids spat out.

To make cakes for storage, clean cloth was laid atop baskets, and a layer of meal was spread on it, sprinkled with water, then covered with another layer of meal. This process was repeated until the cake was built up. The cloth was then tied over the top, and the bundle was set out to dry overnight. A mesquite cake measured about eight inches in diameter, four inches thick, and weighed roughly three pounds when dried. Four or five cakes were made at a time, or, for an extended trip, as many as ten.

Mesquite meal was never cooked, though occasionally a few mesquite pods were added to a pot of cooking squash to add flavor. The Maricopa and some Cocopah cooked cornmeal in water sweetened by mesquite meal.

The Cocopah stored mesquite pods for use in winter and spring in large, circular, bird's-nest-weave baskets, usually placed on a specially built platform five or six feet above the ground. The pods were packed as tightly as possible, trampled down by foot, and broken into pieces as a result. This tight packing was thought to reduce insect infestation.

Comanche – The pods were crushed with long, green mesquite wood pestles in a mortar formed from a hole in the ground lined with a bison hide. The beans were removed, and the remaining mesocarp was pounded into a fine meal, which was eaten alone or mixed with cornmeal to add sweetness.

Mesquite pod meal was mixed with marrow to make a mush or combined with pounded lean meat to make pemmican. It was also mixed with cornmeal in water for a drink, added to soup, or used as a sweetener.

The pods were fed to horses to fatten them. Ripe pods were sometimes boiled, and the beans removed. The seedless pods were then crushed, and the juice strained off. Sugar, and sometimes cornmeal, was added to the juice, which was either drunk or boiled to make a jelly.

Isleta – The pods were roasted and eaten as a confection, or ground into a meal used for making bread.

Kamia – To gather the pods, the Kamia shook trees and used hooked sticks to bend the branches to reach them. They used a cottonwood mortar with long stone pestle to pound up the pods. Kamia mesquite pod granaries were constructed similarly to those of the Cahuilla.

Kawaiisu – The Kawaiisu differed in reports on how the mesquite pods were used by them as far as whether the beans, pods (mesocarp), or both were eaten, but the preparation and storage methods described were similar to other tribes.

Kiowa – Mesquite pods were pounded up and eaten. They pounded the pods into a coarse meal, added sugar, mixed it with water, allowed it to ferment slightly, and then dried it into small cakes. These cakes were sometimes broken down and boiled with meat.

Mexican Kickapoo – Mesquite pods were pounded into a meal.

Paiute – Mesquite pods were a primary wild plant food of the Southern

Paiute. Among some tribes in the Mojave Desert, particular mesquite groves were considered to be owned by families.

Fully grown, but still-green pods were pounded into a pulp in stone mortars with stone pestles and mixed with water to make a beverage.

Dry mesquite meal was made from fully ripened pods. The pods were laid out to remove any remaining moisture, then ground whole into a fine powder. The meal was sifted in an open-twined tray basket to remove the endocarp and seeds, which were too hard to be ground. Finally, the meal was formed into cakes for storage.

The Moapa Southern Paiute made these cakes either in conical burden baskets or in small holes in the ground that were lined with mesquite pod pulp. The baskets or holes were filled with the meal, then after sitting for several days, inverted or dug out, leaving the cones or masses of meal to dry further. Chunks of the cakes were broken off and eaten as-is or added to water for a beverage. The Moapa also mixed dried mesquite meal with cooked agave, forming this mixture into small cakes.

Dried pods were often stored and processed later, but prolonged storage significantly increased rates of insect damage. The dried meal cakes were kept in grass- or bark-lined pits in rockshelters, caves, or underground pits on bluffs and ridges, being areas where very little to no moisture could penetrate.

Pima – The pods were pounded in a mortar, the meal was put through a sieve to remove beans and fibers, and this meal was molded into round cakes and cooked into a sort of bread.

The Gila River Pima in 1775 made the mesquite pod meal into piñole, combining it with tornillo pods, grass seeds, and "other coarse things."

Pueblo – Mesquite bean cakes were stored in large quantities year-round by Puebloans of the Sonoran Desert in 1540.

Seri – The Seri had specific names for eight different stages of growth of the mesquite pods, and different stages may have been prepared differently. The most-used stage was the ripe pods that dried on the tree and fell. The dry pods were gathered from mid-June to late July. Dry pods were also gathered later in the year from pack rat (*Neotoma*) nests.

Dry ripe pods were gathered from the ground beneath the tree. They were usually toasted before pounding them into meal. They were toasted by clearing a section of ground, building a fire there, removing the coals, then placing the pods on the heated ground. At the same time, hot sand, taken from sandpiles heated with a fire atop them, was sprinkled over the pods to toast them evenly. The month in which the pods were processed was called *icóozlajc iizax* – "to-sprinkle moon," the name being derived from sprinkling hot sand over mesquite pods.

The toasted pods were pounded in a mortar of bedrock or hard earth. They used a mesquite wood pestle about three feet long and three inches in diameter. The pounded pods were winnowed to separate pieces of the beans, fiber, and skin. The remainder was further ground into meal and the process was repeated as necessary to remove all remnants of bean, fiber, and skin, leaving just the powdered mesocarp. This meal was cooked as a gruel or thin porridge. For storage, the meal was moistened, formed into cakes or long, thick rolls, dried, and kept in pottery jars. Three individuals could process approximately 88 pounds of mesquite pod meal in a single day.

The fibrous remains winnowed out during the first pounding were chewed to extract their sweet juice, with the remaining pulp being discarded. Sometimes, this discarded pulp was mixed with water to make a beverage.

Sometimes, the coarse meal of the dried whole pods was placed in a pot with water, weighed down with a stone, and left to steep until the infusion was sweet. Dried whole pods were also cooked in water, then chewed, with the fiber and beans spat out, and the decoction was drunk. If the beans were accidentally eaten, they caused a kind of indigestion. Full-size green pods or ripe pods were mashed and then cooked in water. Young, unripe pods were tied into small bundles and cooked with meat.

Shoshone – The pods were pounded in a stone, mesquite, or cottonwood trunk mortar with a stone pestle, and the inedible beans were separated from the pulp. The coarser meal was sifted out, while the finer meal was kept to be eaten uncooked.

Timbasha – The Timbasha Shoshone considered mesquite pods to be as important as pinyon pine nuts in their food system, both of which were their most-eaten wild food products. They pounded the mature, dry pods in a wooden mortar, sifted out the meal, and formed it into small cakes.

At the early stage of growth, when the pods were still green and flat, the pods were cooked in earth ovens, but this did not produce the best-tasting product. At later stages, but while they were still green, the pods were eaten raw as snacks. Different trees were sampled to find those with the sweetest pods.

The ripe pods, at the stage where they were yellow and dropping from the tree but still moist, were pounded in tree stump mortars using stone pestles. The pulp was mixed with water and squeezed out for a beverage. They strongly believed that one should not use both a stone mortar and stone pestle, but one or the other may be of stone.

Dry mesquite meal was made from fully ripened pods. They were laid out to remove remaining moisture, then ground whole into a fine powder. The meal was sifted in an open-twined tray basket to remove the beans (endocarp and seeds), which were not ground or eaten.

The meal was formed into cakes for storage. The cakes were made by first lining a winnowing tray with fibers left over from pounding, then layering the meal on top of these, sprinkling water intermittently to help it adhere. These cakes could reach a foot or more in height and were covered with more leftover fibers that were moistened to form a hard crust. They were then sun-dried for storage. Chunks of the cakes were broken off and eaten without further preparation, or they were added to water for a beverage. The cakes were stored in earthen pits that were lined with grass.

The Timbasha preferred processing the pods before the bruchid beetle larvae emerged, as they were considered to increase the value of the food. The beetles are frequently present in the seeds and mesocarp. They pupate within the pod and later emerge, leaving a small hole. The pods were cached in the summer by burying them in pits that were lined with arrowweed (*Pluchea sericea*), then dug up and processed in the fall when the tribe returned to the area.

Yaqui – The Yaqui used a wooden mortar to grind the pods into a meal and mixed it with water to make *atole*, or a kind of gruel. The meal was also mixed with additional water for a sweet, pleasantly flavored beverage.

Yavapai – The Yavapai pounded up the pods in a bedrock mortar using a stone pestle. One method of consumption was to moisten the meal, suck out the sweet juice, and spit out the solids. The meal was also placed in a watertight basket, moistened, and the juice was pressed by hand into another container and drunk.

Yuma – The pods were made into a mush or bread. They pounded the pods in wooden mortars, mixed the meal with water, kneaded it into a mass, and sun-dried it. They used a mesquite wood mortar, with the bowl burned into it and secured in the ground. In this mortar, the pods were pounded very finely using a stone pestle sixteen to eighteen inches long and weighing about twenty pounds. The beans were separated, and the remaining meal was dampened and formed into large, sweet cakes that were dried. Such cakes could be stored indefinitely in the house.

Sometimes, a basketry mortar was placed in a hole in the ground, and the mesquite pods were pounded into a meal using a stone or wooden pestle. The meal was "pressed into an earthen vessel and set aside to dry," where it could be kept until the next season.

They also soaked the mesquite pods in water and buried them in the ground for two or three days. Once it was almost solidified, the sticky mass was removed and stacked into piles in the shade. This processing lasted several days while the harvest continued, and at the end, the Yuma all feasted on it. This last account about "mesquite," from Forde (1931), may refer to their use of screwbean mesquite (*Strombocarpa pubescens*), based on its similarity to ethnographic accounts of Indigenous peoples of the Southwest processing screwbeans. But it is possible honey mesquite pods were also treated this way.

Mesquite pods were stored in large granaries on platforms of cottonwood poles. The basketry granaries measured three to five feet in diameter and two to four feet deep. The mesquite pods were broken up to pack them more tightly. When filled, the top was covered with arrowweed and mud daub.

Unspecified – In the 1530s, mesquite pod meal was eaten by Texas Natives. It was eaten with "earth" to make it sweeter than its naturally somewhat bitter state on the tree. They dug a hole in the ground, put the pods in it, and pounded them into a meal with a pestle as thick as a leg and nine feet long. Several handfuls of earth were mixed with the meal, and the pounding continued. The resulting meal was put in a small round vessel, and enough water was added to cover the top. The beverage was tasted, and if it was not sweet enough, more earth was added until it reached a satisfactory taste. Thereupon, everyone drank out of the vessel, using their hands as cups. The plant matter was set aside and subjected to three to four washes of water, being squeezed out each time. This was considered a great banquet. In this account by Cabeza de Vaca, it is possible that clay was being used to bind the bitter compounds of mesquite pods. Similarly, Round Valley Natives added clay to acorn cakes to reduce bitterness, and a certain type of clay was used by the Hopi to reduce the bitterness of *Solanum jamesii* tubers.

Many Texas Natives during the Spanish Mission period formed mesquite meal into cakes, or *mezquitamales*, which were sweet and nutritious. After drying the mature pods, they pounded them thoroughly to remove the beans and reduced the remainder to a meal, which was then used to form the cakes.

Southwest Natives bruised the fresh, ripe pods in a wooden or stone mortar, then put them in a pottery container with water and allowed the mixture to stand for a few hours, producing a mush that was eaten by hand. The pods were dried and stored in basketry granaries. The dry pods were powdered, and the meal was moistened to press it into large, thick cakes that were sun-dried for storage. Bruchid beetles in the pods were disregarded and became part of the mesquite meal. The meal was mixed with water for a drink or was made into a gruel.

Indigenous people of northwest Mexico in the 1500s made cakes of mesquite pod meal that could be preserved for a year.

Natives in Santa Fe in 1645 mixed mesquite pod meal with water and drank this mixture as a regular tasty beverage.

Fermentation – The pods were sometimes made into a lightly fermented drink. The pod meal was allowed to soak in water and left to ferment for several days. Sometimes, it was boiled before fermenting (Yuma, other Natives), or other ingredients were added (Mayo). Such fermented mesquite drinks were made by the Cáhita, Cahuilla, Kiowa, Mayo, Opata, Pima, Seri, Tohono O'odham, Yuma, Natives in Santa Fe, and Southwest Natives in general.

Seeds – The endocarp is the outer part of what is commonly called the "bean." Inside this bean is a seed that can be removed. The endocarp or "shell" that surrounds the seed is tough and inedible and must be removed. Most tribes ate only the spongy, sweet tissue of the pod (mesocarp) with the beans (endocarp and seeds) being removed. The "mesquite drink" methods, in which the whole pod meal was soaked in water and strained to make a beverage, were likely common because they avoided the need to remove the beans.

The seeds were, however, eaten by the Apache, Chemehuevi, Moapa, and Seri. Considerable work was required to process them. The beans were parched, pounded to remove the "shell" (endocarp), and the seeds were ground into a meal using a metate (Chemehuevi, Moapa). Sometimes, raw beans were ground on a metate, with the shells winnowed off (Apache). The seed meal was made into bread or pancakes (Apache). The seeds were often cooked with meat (Apache).

The Seri separated the beans from the pods and ate only the mesocarp (the fibrous exocarp or skin was also largely removed) as their principal mesquite pod meal. However, they also sometimes separated the seed from its shell, ground the seeds into a meal, mixed it with water and mesocarp meal, and drank the resulting mixture.

The difficulty in separating the seeds from the inedible shell (endocarp) was why they were not more commonly eaten by Southwest Natives. Prehistoric Natives of Sonora appear to have used a special type of mortar that was excellent for crushing the endocarp to extract the seeds, as well as for grinding the pods (mesocarp). These mortars were toroidal (donut-shaped), and a bulbous-tipped pestle was gyrated around the hole as an axis.

Flowers – The flower spikes (racemes) were eaten by the Chiricahua Apache, Cahuilla, and Cocopah. Flowers vary in their sweetness, and the better ones would be put in a pan or bowl of water and gently mashed. The resulting infusion was drunk, leaving the flowers behind to repeat the process with two or three more washes, after which the flowers themselves were eaten (Cocopah).

These flowers were roasted in an earth oven, then "squeezed into balls ready for eating" (Cahuilla). These balls, known as *selkulat* (i.e., "blossoms made of"), were stored in pottery vessels and cooked as needed in boiling water. The Cahuilla also drank mesquite blossom tea.

Sap – Honey mesquite sap was eaten by the Pima, Seri, other Southwest Natives, and Mexican folk. The sap, or its hardened, gummy resin, was dissolved in water to make a drink. The gum was also chewed, spat out, and followed by a drink of water, which tasted sweet due to the gum (Seri). The sap or gum from the tree is similar to gum arabic.

Roots – The roots were used to flavor *tiswin*, a fermented drink of the Apache. As a flavoring for fermented agave, Berlandier (Ohlendorf et al. 1980) notes the use of a bitter root called "*raicilla*" described as a "mimosa" (Fabaceae) and also called "*mezquitito*." This was likely the mesquite root, as Castetter and Opler (1936) recorded its use by the Apache for the same purpose.

Notes

Character – In my research, this was the single most important wild plant food among Indigenous peoples of the Southwest historically. The reason is not only that it is a common tree throughout the region, but also that its pods are produced in great abundance. There are three reasons for this profusion: 1) It has deeper taproots than any other plant, allowing it to thrive in arid regions. 2) It fixes atmospheric nitrogen in its roots, enabling more efficient growth. 3) Its intended animal dispersal agent is likely extinct, leaving the pods to be dispersed primarily by humans, livestock, and feral pigs. The seed bank built by dropped pods also assists rebound.

Honey mesquite leaves are bipinnately compound, with two (rarely four) pinnae coming off each petiole, each with twelve to twenty foliolules (leaflets). The branches have stout, straight thorns up to a few inches long. Its flowers are cream to yellow, plume-like or spike-like racemes. It has rough, reddish, fissured bark. Its fruits are bean pods, four to nine inches long.

Many mesquite beans were infested with bruchid beetles, but these infestations were disregarded by the Cahuilla, Pima, and other Natives. Pods were gathered and processed regardless of infestation. The larvae became part of the meal, so the final product was a mix of insect and plant matter, which likely increased its fat and protein content.

Until about 10,000 years ago, mesquite bean pods were probably dispersed by the extinct, elephant-like gomphotheres.[33] Today, cattle and horses aid in its dispersal.[33] The thorns and exceptionally hard, resilient wood of mesquite may be adaptation to gomphothere browsing, helping to deter excessive breakage of the branches.

There is a popular myth that mesquite is an invasive species introduced by the Spanish, but that has no basis in fact. The USDA and every other plant database list it as native to Texas and the Southwest. The very first Spaniard in Texas, Cabeza de Vaca, described Natives eating mesquite in 1530. Archaeobotanical and genetic studies show mesquite has been in the Southwest for about a million years. It is invasive in parts of Africa, where it was introduced. This may partly explain the genesis of the myth, combined with the antipathy of Texas ranchers toward the species because they dislike how well it rebounds when they clear-cut land for livestock. This is similarly the origin of the "water-hog" myth also leveled at mesquite (as well as juniper), which is also proven false.

Season – Mesquite flowers are available from March to September, usually peaking around April and May. The full-sized green pods are available from late spring to late summer. The mature pods are available from early summer to fall. The peak gathering season for ripe mesquite pods in Austin is mid-June to mid-July.

Nutrition – The sweet taste of mesquite pods comes from the spongy mesocarp tissue (between the beans and the skin), which contains 87% carbohydrates, 7% protein, and 2% fat.[31]

A drink made with 20 pods (about 60 grams), has approximately 236 Calories, 52 grams of carbohydrate, 4 grams of protein, and 1 gram of fat. Such a drink can be made in about 10 minutes with only a couple rocks and a water bowl. This high ratio of calories consumed to calories expended in preparation is one reason mesquite pods were so highly valued historically.

The seeds contain a trypsin inhibitor (6 TIU/mg), but the mesocarp or pericarp has only minor concentrations (1 TIU/mg).[31] Of the four Indigenous cultures known to eat the seeds, three cooked them, which destroys or reduces the trypsin inhibitor; the Seri, however, were not recorded as cooking them.

Practice – Mesquite pods are produced extremely prolifically, the tree is common throughout Texas, and they are one of the best wild food sources in the state except in further east Texas. I gather them in large quantities, but first sample the trees. Some trees produce small, skinny, bitter, and damaged pods, while others yield large, fat, sweet, and undamaged ones.

I prefer to gather them when they are fully mature. At this stage, they are easy to pull off the tree, and have begun falling. They also have a faint rattling sound when I shake them, indicating that they are dry. I gather the pods by pulling them off the tree by hand, shaking the trees and branches to cause the pods to drop, or by simply picking them off the ground.

One study in Arizona showed high aflatoxin (a carcinogen produced by fungi on crops and foods) concentrations of mesquite pods gathered from the ground.[12] The samples in this study were gathered from October to March, after the seasonal monsoon rains in Arizona, and several months after I would personally want to gather mesquite pods from the ground in Texas as they clearly look bad, developing a dark color by this time. I prefer to not gather them from the ground when it has rained significantly since they fell. If the pods have had time to linger in a wet condition long enough for appreciable fungal growth, they are unfit for consumption. Even in dry conditions, I generally do not gather pods from the ground after it has been more than a few weeks since they fell.

I have made "mesquite drink" countless times. It tastes sweet, with a flavor like caramel tamarind. I pound the pods into a coarse meal in a large wooden mortar and pestle made from a black willow trunk and a heavy mesquite branch. The pods must be fully dried before pounding, otherwise they will form a sticky mass that is difficult to reduce by pestle. I soak the coarse pod meal in water, then strain it to obtain mesquite drink. The meal needs little soaking time; the sugars are extracted almost immediately. Large batches may benefit from a second, longer soaking, and even a third, in order to extract every bit of nutrition.

The easiest way I make mesquite drink is to fill a blender with the whole pods, add enough water to cover them, blend the mixture, and strain it through wire mesh. Mesquite drink is excellent plain but also has a flavor profile complementary to coffee, cocoa, cinnamon, cardamom, clove, nutmeg, and ginger (many "pumpkin spice" flavors).

I have made the meal by pounding the dry pods into a coarse meal in a large mortar and pestle, straining out the beans with a coarsely woven basket, then sifting out the finer meal by more finely woven baskets and tossing the meal in a flat tray to separate off the larger particles.

The easiest way I make mesquite flour is also in a blender. To blend the pods into a meal, they need to be very dry; otherwise, they will form a sticky mass. I sometimes give them a light cooking in the oven, just enough to dry them thoroughly. I usually need to stir the pods as they are blended to grind them completely. A finer-grade wire mesh sieve, such as one used for flour, will yield a superior product, but the coarse meal sifted through a wire mesh strainer is adequate. Mesquite meal can be added to pretty almost any baked good for excellent flavor. Moistened meal formed into cakes and dried also tastes great.

When storing the pods at ambient temperature, insect larvae (mostly bruchid beetles) will develop, emerge, and re-infest the pods. They can still be used, but the adult insects are a nuisance if the pods are kept indoors. I keep them in a dry, sealed container for several months to a year, but they are best processed into a meal immediately after harvest and drying. The most important factor in storage is keeping the pods as dry as possible. If they become moistened, which merely daily temperature variation can cause via condensation, they may grow mold and become unusable.

Mesquite seeds are very hard to grind, requiring a stone mortar and pestle (or a mechanical device). The beans (shells) are often split open in the process of pounding the pods, releasing these small, brown seeds. The raw ground seed meal tastes mild and nutty, but is best cooked.

The sap has a mildly sweet, somewhat resinous flavor. A hot infusion quickly dissolves the sap to make a drink.

The flowers vary in taste, ranging from bitter, astringent, and insipid to mildly astringent, herbaceous, and faintly sweet. They may be sweeter when freshly bloomed and still containing nectar. I have eaten many whole and raw, but they also make a decent herbal tea.

Neltuma glandulosa. Base photo: full-sized tree. Overlay, clockwise from top left: inflorescence; flowers in early April; range map (GBIF); bipinnate (two-tiered) compound leaves; harvested mature pods in early July; close-up of pod cross-section; reddish, nearly-mature pods in mid-June; long, robust, straight thorns.
Pod cross-section labels: a) exocarp (skin); b) mesocarp (food); c) endocarp ("shell"); d) seed coat, e) endosperm; a-c) pericarp; d & e) seed; c-e) "bean."

Parkinsonia aculeata L.
Paloverde

Jerusalem thorn, Barbados flowerfence, horsebean, *retama, bagote, lluvia de oro*
Seri: *snapxöl*
Loc.: W, S, SE, & C TX; common in Travis Co.; 2 *Parkinsonia* spp. in TX.
Form: shrub, tree, up to 40 ft. tall; perennial.
Flowers: Apr-Aug (yellow, orange).

History

Fruits – The Tohono O'odham at the beans of this species. The pods were threshed to release the beans, which were then winnowed and parched.

The Seri ate the beans of two other *Parkinsonia* species, but not those of *P. aculeata*. Since *P. aculeata* was only cultivated or semi-naturalized in Seri territory, it may have lacked the abundance, familiarity, or history needed for food use like the other two species.

The Cahuilla and Seri ate the beans of *P. florida*, while the Tohono O'odham and Seri ate those of *P. microphylla*. Among the Seri, green pods were eaten cooked with meat. The Cahuilla ground mature, dry beans into a meal that they cooked as a mush. The Seri and Tohono O'odham sometimes toasted the beans before grinding.

Notes

Character – The name "paloverde" (Spanish for "green tree") refers to the tree's distinctive green bark. It bears thorns and long, linear leaves that have dozens of tiny leaflets attached, in a pinnately compound arrangement. These leaflets fall off in response to aridity (or grow after rains), leaving only the green rachis (linear leaf-like structure). The green bean pods are similar in size, shape, and taste to soybeans.

Parkinsonia texana is also found in Texas (SW & S TX).

Season – The green beans are available from late spring to midsummer. The mature, brown and dry beans are available from early to late summer.

Nutrition – The dry beans contain 6% moisture, 4% ash, 16% protein, 1% fat, and 28% fiber.[29]

Practice – I strongly prefer using the green beans for food. The green beans can be stripped out of the pods with hands or teeth and eaten fresh and raw, and taste like soy beans. They taste even better after the whole green pods are boiled, salted, and eaten like edamame.

The mature, dry beans are very hard to grind. I have to use a stone mortar and pestle and a decent amount of force. A grain mill would probably also grind the beans. Once ground, the bean meal must be cooked. When the bean meal is prepared and seasoned like pinto beans, it tastes quite good.

Parkinsonia aculeata. Base photo: fruiting tree in late July. Overlay, clockwise from top left: range map (GBIF); pinnately compound leaves (sometimes lacking leaflets) and thorns; harvested, threshed, and winnowed mature, dry beans in late July; harvested young and tender bean pods in late June; green bark with thorns.

FAGACEAE – Beech Family

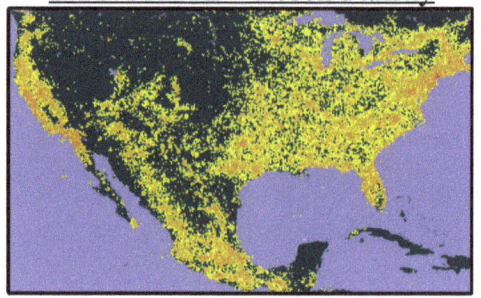

Quercus distribution. Source: GBIF.

Quercus spp.
Oak

Apache: *tcintcile*, Cherokee: *táᵈłá*, Comanche: *pasapɔni / pasaʔ ponii* (acorns) / ‹*makine*› / *tuhuupit* / ‹*pibi*› *tuhuupit* / *pibiahuupi* – "real oak," Diné: *tchétc'il* – "rock plant," Gosiute: *ku'-ni'-ûp*, Haudenosaunee: *ogowä'*, Hopi: *kwi:'ngvi*, Tewa: *kwæ*
Loc.: all TX; very common in Travis Co. (11+ spp.); 50 spp. in TX (7 cultivated), plus 46 hybrids.
Notes – This is the 5th most speciose genus of vascular plants in Texas.[35]

History

Fruits – Acorns were eaten by the Chiricahua, Mescalero, and Western Apache, Asinai, Caddo, Cahuilla, Calpella, Cherokee, Northern Cheyenne, Choctaw, Comanche, Concow, Dakota, Diné, Gosiute, Haudenosaunee, Hocąk, Hoopa, Isleta, Karuk, Kawaiisu, Kichai, Mexican Kickapoo, Kiowa, Little Lake, Maidu, Menominee, Meskwaki, Miwok, Nabedache, Nomlaki, Ojibwe, Omaha, Pawnee, Pomo, Ponca, Potawatomi, Salteaux, Sauk, Shasta, Tohono O'odham, Tewa, Tolowa, Wailaki, Yokia, Yokuts, Yuki, Yurok, Great Lakes Natives, Mojave Desert Natives, California Natives, Natives in Arkansas, South Carolina, and Florida in the mid-1500s, Puebloans of Arizona and New Mexico in 1540, Southeast Natives, and American Indians in general. Species eaten include *Quercus agrifolia*, *Q. alba*, *Q. chrysolepis*, *Q. douglasii*, *Q. dumosa*, *Q. ellipsoidalis*, *Q. emoryi*, *Q. gambelii*, *Q. garryana*, *Q. grisea*, *Q. kelloggii*, *Q. lobata*, *Q. macrocarpa*, *Q. marilandica*, *Q. michauxii*, *Q. muehlenbergii*, *Q. nigra*, *Q. oblongifolia*, *Q. prinoides*, *Q. rubra*, *Q. sadleriana*, *Q. stellata*, *Q. turbinella*, *Q. velutina*, *Q. virginiana*, *Q. wislizeni*, and *Q.* x *pauciloba*.
The California way – Native peoples of California relied heavily on acorns, comprising as much as half of all calories consumed. Their methods of acorn processing were well documented during the historical period. These peoples include the Calpella, Cahuilla, Concow, Karuk, Kawaiisu, Maidu, Miwok, Nomlaki, Pomo, Shasta, Tolowa, Yokia, Yuki, and Yurok.

The methods used by these peoples are fairly similar, though many variations existed. Generally, acorns were gathered in large amounts, dried, then hulled to remove the shells. The kernel skins were rubbed and winnowed off. The kernels were ground into a fine meal, which was leached with water. Leaching basins were built, into which the acorn meal was placed and water was poured to soak the meal and allow it to percolate out. The leaching continued until the meal no longer tasted bitter. Then, the meal was suitable to eat, plain, cooked as mush, or baked into cakes. The specific methods of these Natives, from gathering to cooking, are detailed below. After that, other methods are listed.

Gathering – Acorns from different species were used to varying degrees in different areas, since multiple oak species were found in most areas, all of which were usable for food. One species was typically preferred based on acorn size, tannin concentration, and the amount of good acorns that could be gathered.

Acorns were gathered by beating branches, using long poles to knock them down, climbing trees and striking fruiting branches with sticks, cutting down small branches, or simply picking them up from the ground. Among the various reasons Natives in California frequently practiced low-level burning was to kill acorn weevils, reducing infestation rates, and to make acorn gathering easier by clearing debris from the forest floor. Acorns were placed in large burden baskets, such as conical baskets carried with a tumpline across the forehead, which freed both hands for picking and tossing acorns into the basket. Up to eight large basketfuls, or about 500 pounds, were gathered for the year's supply.

Drying – Acorns were initially sun-dried. This made hulling easier, if they were to be processed immediately, and prevented mold growth if they were to be stored in a granary. The Kawaiisu cracked each acorn, but did not remove the hull, allowing them to sun-dry for a couple of days, which made the kernels easier to remove when they were finally hulled.

Storage – Acorns were stored in small, elevated granaries in trees or on posts. Such granaries were woven from coarse withes and thatched or covered with hides or bark. Kawaiisu granaries were built from hardwood poles with the sides, top, and bottom covered with bark and lined with *Eriogonum* sp. leaves, and elevated about a foot off the ground. Acorns were also stored in stone-lined pits covered with brush, large flat stones, and bark, or in caves.

For long periods, acorns were generally stored with the hulls on. Granaries were sometimes filled with hulled acorns, but only if they were still drying and would be used soon. Care was taken to prevent mold growth. After about a year, acorns were shelled and the meat was ground up and leached with cold to hot water. The resulting meal was eaten or dried for preservation. Acorn meal and dried cakes were also stored. The Caddo and Nabedache in Texas similarly stored acorns in great quantities in basketry granaries.

Hulling – After sun drying, further drying, and storage in granaries, the acorns were cracked with the teeth or a small stone and hulled. Acorns were placed on an anvil rock with a small pecked indentation, in which they were struck with a hammerstone to remove the hulls. The anvil rock was sometimes simply a flat stone. The Karuk hulled acorns as they gathered them, cracking them in their teeth and filling their baskets with only the kernels. The thin membrane covering the nut meat (kernel skin) was rubbed off by hand, sometimes aided by wetting them or a handful of grass. The kernel skins were then winnowed off and the kernels were spread out to dry.

Grinding – The dry kernels (nut meat) were pulverized in a bedrock mortar or on a large flat stone. A shallow, bottomless hopper basket was sometimes attached to the mortar's rim with pitch or asphaltum, or simply held in place by hand. This hopper caught flying pieces of kernels as the stone pestle struck them. Without a hopper basket, the meal was simply swept back into a pile with one hand after each strike while the other held the pestle. Manos and metates or portable mortars and pestles, made of stone or mesquite wood, were also used. Stone pestles used by the Shasta for pounding acorns were about eight to ten inches long.

Periodically, the finer meal was separated from coarser particles by spreading it evenly on a circular, flat, finely-woven basket and shaking it vigorously so that the larger particles fell into a basket held below. Meal that clung to the basket after several rounds of spreading, shaking, and tapping with a stick was very fine. This finest-grade meal was then knocked loose from this basket with sharp blows from a bone, releasing it into a receptacle below.

The coarse meal was pounded further, with additional acorns periodically added to keep the pile at a consistent mass. The process continued until all the meal reached the finest possible particle size. Pieces that fell outside the lower collecting basket were carefully swept up with a brush and returned to the pile for grinding. Finely ground kernels facilitated both tannin leaching and the nutrient absorption. The grinding process was time-consuming, and having the finest acorn meal was considered socially desirable.

Leaching – To leach the tannins from the meal, a sand basin was often used. The meal was mixed with some water and poured into a shallow depression dug into sandy soil that was often lined with conifer needles. The depression was circular, with a diameter of about one to three feet and depth of about two to three inches, with the earth being heaped up in a little wall around the circumference. The Kawaiisu dug these basins in regular soil, lined them with willow leaves, then packed a layer of clean sand on the bottom.

Water was carefully poured over the meal, which was often covered by conifer needle sprays to make the flow of water more gentle and prevent sand from stirring into the meal. The needles also imparted a slight piney flavor. For several hours, or until the meal no longer tasted bitter, water was periodically added as it slowly percolated through meal and sand, leaching the bitter tannins. If sand was unavailable, dry pine needles could be arranged to form the basin and a coarse cloth kept the needles from sticking to the meal.

For a leaching basin, the Cahuilla used a basket filled with sand or a coarsely woven basket lined with grass, leaves, or fibrous material. The Shasta constructed a platform of sticks six to twelve inches off the ground, supported by stones or forked sticks planted in the ground. The platform was covered with a thin layer of pine needles, followed by a layer of sand, which was made thicker at the edges to form a basin. The meal was spread in the basin about two inches thick, and warm water was poured over it to leach the tannins.

Cool water was often used for the whole leaching process, but warm or hot water was sometimes employed. The Northern Maidu used either cool water (fresh from the source) or warm to boiling water heated in baskets with hot stones. Hot water leached tannins faster but decreased the quality of the meal. The Northern Maidu began with warm water for the first soaking, gradually increasing the temperature with each successive soak, and finished with boiling water. Generally, cool water was not used after warm water, except that the Kawaiisu alternated between hot and cold. The Karuk started with cool water and gradually increased the temperature to remove all the tannins. The Cahuilla also sometimes used warm water to speed the process. After leaching, the meal was tasted to ensure it was no longer bitter; if necessary, soaking continued.

Cooking – Once the meal was thoroughly leached, most of it was scooped out of the basin to be cooked into bread. The Shasta removed it from the leaching basin by slapping a hand on the wet meal, which adhered to the hand, which was then dipped into a basket full of warm water to carefully wash off the sand. The meal remaining on the hand was put in another basket of water for cooking or dried for storage. The meal clinging to the sand was made into a soup since the sand settled to the bottom of the cooking vessel and this portion was left unused.

The usual proportion for the Northern Maidu was about two quarts of meal to three gallons of water, or less water to make a thicker mush. The mixture was stirred and brought to a boil in watertight baskets with hot stones held by tongs of green wood. The hot stones preferred by the Shasta were porphyritic or close-grained igneous rocks. Sometimes, a leaf or two of mint or bay laurel was added to the soup for flavor. Such soup, mush, or gruel was a staple food. It was eaten plain or with piñole or venison, using the fingers or a mussel shell spoon as utensils. Coarse meal was cooked into mush by the Cahuilla. Chia seeds, wild berries, or meat were sometimes added. This mush was allowed to cool and jell into a firm consistency, then was cut into cubes for eating. The Kawaiisu cooked the mush, then let it stand and harden into a kind of cake that was a staple food usually eaten with meat, especially deer meat.

The Northern Maidu made a loaf about six inches in diameter that was flattened, then a hot rock rolled in oak leaves was placed in the center, and the dough was folded over and pressed down all around it. The resulting bread was solid, heavy, lumpy, and almost tasteless. Before being baked with clay, a small amount of tannins could remain, imparting a slight bitter or astringent taste to the meal. In one method to make bread used by Natives in Northern California, the acorn meal was mixed with $1/20^{th}$ part of red clay. It was cooked in an earth oven overnight, or about twelve hours, before being removed. The bread turned dark black on account of the red clay reacting with the remaining tannins, and it hardened as it cooled. This bread contained only trace tannins and was therefore sweeter than the mush prepared without clay. The Shasta made small cakes from the meal, baking them on flat rocks slanted in front of the fire and eating them with salt. The Cahuilla made cakes from the fine meal that were baked in hot coals for several hours.

Burial – A simpler, though often less practical, method of tannin removal (leaching) was to place acorns into wet clay for about a year (Northern California Natives). The Shasta buried whole acorns of *Q. chrysolepis* in the mud for several weeks until they turned black. The acorns were then dug up, cracked open, and the kernels were boiled whole without first grinding them into a meal.

Occasionally, Round Valley Natives buried acorns in sandy soil along with grass, charcoal, and ashes. The acorns were periodically soaked with water until they became sweet. The Ojibwe gathered *Q. macrocarpa* acorns in the fall, buried them, and ate them in the winter, spring, or early summer.

Boiling & ashes – Acorns were often boiled with wood ashes to leach out the tannins (Dakota, Haudenosaunee, Mexican Kickapoo, Menominee, Meskwaki, Ojibwe, Pawnee, Ponca). The ashes were likely from hardwoods (conifers have less potassium), essentially creating a lye solution (potassium hydroxide) that helped remove the tannins. The strained decoction of wood ashes is lye, but it is unclear in most accounts whether the ashes were strained out. The effect is the same, and there is no need to strain the ashes, since the acorns would need to be washed off either way. Some accounts mention no washing stage, implying acorns were eaten with lye residue, which is not recommended. Lye saponifies lipids (breaks down fats), so is used for making soap. Lye, especially in stronger solutions, is caustic, and can cause chemical burns on the skin and blindness if splashed in the eyes.

The Ojibwe made lye by boiling wood ashes and soaking the (likely hulled) acorns in the decoction. They put these lye-soaked acorns in a woven bark bag and washed out the lye with several changes of warm water. The washed, leached acorns were dried for storage and ground into a coarse meal that was cooked as mush or in soups.

In a Menominee method, *Q. alba* acorns were parched, their hulls threshed off, and the kernels boiled until nearly cooked. Then, the water was discarded, fresh water and two cups of wood ashes were added, the mix was boiled, and the acorns were strained out. The acorns were then simmered in a third change of water to clean them. They were then ground into a meal (likely after being dried), sifted, and eaten in soups or as mush.

Hulled acorns were apparently also simply boiled (Comanche, Diné, Ojibwe), sometimes in multiple changes of water (Haudenosaunee). Those accounts may have neglected to record the addition of ashes, but it is possible to leach whole acorns to some extent simply by boiling them, especially if the water is changed.

Roasting – Acorns were occasionally roasted and eaten without leaching, sometimes being ground into a meal afterward (Diné, Ojibwe, Shasta). The roasted acorns of *Q. alba* were made into an infusion or decoction that was drunk by the Meskwaki as a beverage. The Gosiute and Kiowa made a similar beverage from *Q. nigra* or *Q. stellata*.

Raw – Some acorns, including from *Q. emoryi*, *Q. grisea*, *Q. gambelii*, and *Q. marilandica*, were eaten raw (Chiricahua, Mescalero, and Western Apache, Comanche, Tohono O'odham).

Other ways – Acorns were made into a kind of pemmican by the Chiricahua and Mescalero Apache. They gathered ripe acorns, lightly roasted them, pounded them up, mixed them with dried meat or fat, and stored them in hide containers that kept them throughout the winter. This served as a high-quality emergency food for long-distance trips.

The Choctaw ground acorns into a meal and leached it in a cane basket sieve by periodically putting water in it. The meal was cooked as a mush, used like cornmeal, added to stew, or molded onto the ends of sticks and cooked over a fire.

Importance – Acorns were historically the most important wild plant food among Indigenous North Americans in general. They also feature prominently in the diets of Indigenous peoples in Europe and East Asia, or essentially wherever *Quercus* species occur.

Acorns were gathered and stored in great quantities, usually enough to last at least a year (Caddo, Nabedache, Natives in Arkansas, South Carolina, and Florida in the mid-1500s, California Natives, Southeast Natives). Acorns were also an important food source of the Menominee, Salteaux, Sauk, Potawatomi, and tribes of the Great Lakes. Acorn season was one of the six periods used to divide the year by the Chiricahua Apache. The Asinai said that a woman who sprang from an acorn created the outlines of heaven by placing timbers in a circle. Acorns have been eaten in North America for at least 10,000 years. In general, about half of the Californian Native diet was acorns.

Various species of oaks can produce from 160 to 500 pounds of acorns per year. A single stand of oaks could adequately satisfy an entire village's demand for acorn meal. It is estimated that an average family, working normal hours for two weeks, could collect over 33,000 pounds of acorns. Species of oaks vary in how abundantly they bear crops each year. Some produce abundantly every other year, or one in three or four years. Different species of oak have different protein, fiber, and tannin contents, influencing taste, digestibility, and rate of decay, shaping the preferences of Indigenous peoples.

The average acorn cache of the Sierra Miwok stored enough acorns to produce 50 grams of protein per day per person, exceeding the average male requirement of 37 grams. Acorn meal, pinyon nuts, mesquite pods, and palm dates were the main trade items of the Cahuilla. Acorns were considered a very fattening food. Groves or areas of especially fruitful oaks were often owned and cared for by a particular tribe or village, and individual groves and trees were often owned and cared for by a particular family or individual. The number of such oaks or other highly used trees and shrubs owned by the Pomo represented a measure of wealth.

Acorn meal has a similar nutritional profile to mesquite pod meal, with slightly higher fat content and slightly lower carbohydrates. Mesquite pod meal is easier to prepare, requiring only grinding and sifting, without hulling, leaching, or cooking. Mesquite pods also ripen earlier in the year, in midsummer, compared to acorns in the fall. In Austin and the greater Southwest, where both mesquite and oak trees grow, mesquite served as the preferred staple wild plant food, furnishing abundant carbohydrates with minimal effort. While many Natives in the Southwest also ate acorns, they were far less important in the diet compared to regions without mesquite.

Notes

Character – Acorns are generally inedible raw because they have varying concentrations of tannins, a bitter compound that can be harmful in large amounts. While some acorns have low enough tannin levels to be eaten raw, most require processing to remove them.

Season – Acorns are generally a fall crop, but acorns from various species ripen from late summer through November. In Austin, peak acorn season runs from mid-September to mid-October, but fallen acorns can be gathered into November, and mature ones may be pulled from trees as early as late August.

Nutrition – Prepared acorn meal from *Q. rotundifolia* and *Q. ilex* averages 5% protein, 11% fat, and 80% carbohydrate.[57] The meal is rich in monounsaturated (62% of lipids) and polyunsaturated (16%) fatty acids.[57] *Q. virginiana* acorns contain about 6% protein.[26]

Practice – I start looking for productive oak trees in late September. Those of *Q. muehlenbergii* are the best, and I gather them in early to mid-October. Individual oak trees rarely produce the same amount of acorns every year, often following a two-year cycle. The easiest acorns to collect are beneath trees that have cleared ground, such as short grass, dirt, or mulch. They are more difficult to gather on an intact forest floor, where they are obscured by vegetation and leaf litter. Trees overhanging dry creeks also make good acorn collection sites. In Austin, I have harvested, processed, and eaten acorns from *Q. buckleyi*, *Q. fusiformis*, *Q. macrocarpa*, and *Q. muehlenbergii*.

Of the four, *Q. fusiformis* acorns are the poorest choice. *Q. sinuata*, *Q. stellata*, and *Q. virginiana* produce comparable acorns, though I have not processed those for eating. While *Q. fusiformis* is still perfectly usable, especially from trees that yield large acorns, they are usually small, frequently insect-damaged or malformed, and higher in tannins. Despite their abundance, they are my last choice. They are the squirrels' acorns.

Q. buckleyi acorns were likely the most valued acorn in this area by pre-contact Native peoples, though I have found no historical evidence that acorns were used for food in Central Texas. It may be that acorns were not eaten in the area because mesquite already provided sufficient carbohydrates with far less processing. But if acorns were eaten, *Q. buckleyi* would have been a prime choice. *Q. macrocarpa* and *Q. muehlenbergii*, while now common in Austin, were likely absent until they became common landscaping trees. Both are native to Texas but not Travis County. *Q. buckleyi* acorns are abundant and larger than those of other native oaks in Austin. Their kernel skins slip off easily and their tannin concentration is about average for oaks.

Q. macrocarpa acorns are exceptionally large, making them a good food source but also more susceptible to mold, as they do not dry out as easily as smaller acorns. I have lost large amounts of these when they were not fully dried and fungal growth corrupted the kernels. They are somewhat more difficult to hull than other acorns as the acorn cup (the woody "cap") often envelops the shell and must be cracked off first. The kernel skins are easily removed. The huge kernels must be dried a long time, then stored properly. They are also somewhat difficult to grind, as their size allows only one or a few to be processed at a time.

The Ojibwe methods of leaching this species may be good choices. They either soaked these acorns in lye and rinsed them with several changes of water, or simply buried the whole acorns in mud in the fall, then dug them up for eating at any time between winter and early summer. The Northern Cheyenne dealt with this species by roasting or parching them in the hull (which may have served to dry the kernels), hulling and cleaning the kernels, grinding them into meal, and boiling it to eat as mush. Perhaps they changed the boiling water to remove the tannins.

Q. muehlenbergii acorns are the best in Austin that I have tried. They are commonly grown as landscaping plants, so their acorns are often on cleared ground and easy to gather. They are relatively large, easy to dry and hull, and their kernel skins fall off readily. They also have low tannins and a sweet taste, being almost edible raw, so the leaching process is rapid. The following steps can be used for any acorn, but I typically do it for *Q. muehlenbergii*.

In general, I follow the historical methods of Indigenous peoples of Northern California, using Paleolithic technology. This approach provides excellent survival knowledge. Without proper baskets, one can use a tray made from a large section of bark to sift the acorn meal by tapping, at least allowing one to separate large particles from the coarse meal.

1. **Gather & dry**: After gathering acorns from the ground, I give them a quick rinse and then set them out to dry. Depending on how recently they fell from the tree, they usually become completely dry within one to three days. Rinsing is not strictly necessary.
2. **Storage**: If I am not processing the acorns immediately, I store them in paper or cloth bags indoors. For long-term storage, however, I prefer a watertight storage container outdoors, as acorn weevils and moths will emerge and can be a nuisance indoors. Be sure the acorns remain very dry to avoid mold.
3. **Hulling**: I do not believe any modern technology is significantly superior for hulling acorns compared to a traditional hammerstone and flat stone anvil. Fancy machines to hull acorns exist, costing hundreds of dollars, but in my experience, they are not faster than simple rocks.

    I get into a comfortable position, place a larger flat anvil stone in front of me, and use a quartz cobble to crack the hulls open. I always direct the "point" of the acorn upward, placing the end where the cup was attached directly against the anvil stone. This orientation usually allows the hull to split in a single, or occasionally double, strike. Placing the point downward is slightly less optimal, but trying to crack the acorn on its long side often results in a crushed kernel inside an incompletely cracked hull. The optimal hammerstone size depends on the acorn. Huge bur oak acorns require a larger stone. For most acorns, I use a stone about the size of my fist or smaller.

    With a bit of practice, it becomes muscle memory: grasp an acorn with the left thumb and middle finger, place it on the anvil point up, strike it with a measured blow, split the hull in half, drop it in a pile, and place the kernel in a basket, wiping off its skin along the way. For species whose skins adhere to the kernels, it is necessary to detach and winnow them. The skins can usually be detached by vigorously rubbing the kernels between the palms. For strongly adhered skins, it may be necessary to crack the kernels with a stone first, then rub them off.

4. **Grinding**: Making acorn meal is more of a cycle than a linear process. Between each step, I sift the meal, shuttling the coarse meal back one step for re-grinding and moving the fine meal forward as finished. More acorns are also added between each step to replace the volume of finished meal removed.

    First, I place a double handful of kernels in my large wooden mortar and pound them until they break into small pieces. Finer meal produced from this initial pounding is set aside. After this first reduction, I transfer the broken pieces into a stone mortar. As a pestle, I use a natural oblong stone cobble about eight inches long and two-and-a-half inches in diameter. I have tried using a basketry hopper, but I prefer to contain the meal by sweeping with my left hand. The kernels mostly fly out only during the very first stage of grinding whole kernels, which is why I use the deep wooden mortar for this stage.

    Acorns can also be easily ground into meal using a blender, coffee grinder, or food processor.

5. **Sifting**: To separate coarse meal from the larger particles, I shake the meal on a flat basketry tray, allowing the larger particles to fall into a basket below. Coarse meal can be leached and cooked into mush, so this is the only necessary step. I re-grind the larger particles, adding more crushed kernels to replace the finished coarse meal.

    For a superior product, I separate fine meal from the coarse meal and re-grind the coarse portion. For the finest grade, I only use meal that sticks in the weave of a basket. To do this, I put the ground meal into a basket, shake it gently, then pour out the loose meal. I then turn the basket upside down over a flat tray and strike the bottom with a stick, releasing the fine meal trapped in the weave.

    Acorn meal can also be sifted using a wire mesh flour sifter for fine meal or a wire mesh strainer for coarse meal.

6. **Leaching**: I built a sand basin, using coarse sand and gravel for the body and fine, cleaned sand for the bowl where the meal is placed. I slowly pour water over the meal, ensuring it is evenly distributed.

To prevent the meal from being overly agitated, I pour the water over sprigs of juniper needles. After several rounds of soaking and draining with cool water, I taste the meal. I continue soaking until it no longer tastes bitter.

I scoop out the leached meal by hand, taking care with the bottom layer to avoid picking up excess sand. The meal sticks to itself, making it easy to separate, though it will pick up a thin layer of sand grains. I place this bottom portion in a separate container; when water is added, the sand sinks. I just pour out that fraction after eating the meal from the top.

A finely woven cloth bag is an excellent modern leaching tool. It can be hung to gradually drain water, which is replaced intermittently, or put it in a large vessel with occasional water changes.

7. **Cooking**: I build a fire and heat stones (solid, uncracked quartz cobbles gathered from a dry location), placing them directly in the fire and continuing to stoke it. Meanwhile, I place hefty pats of the freshly leached, still-wet acorn meal onto large, fresh or dampened leaves. Among the largest native leaves in Austin are the lower leaves from sycamores (*Platanus occidentalis*), which are non-toxic. Large oak leaves, such as those from *Q. macrocarpa*, are better for flavor. Grape leaves can also be used. Dried leaves should be soaked in water before use for flexibility. I remove the hot rocks from the fire by knotting the end of a small, forked sugarberry (*Celtis laevigata*) branch into a small hoop, then using a second stick to roll the stones into this hoop, shake off the ashes, and place them for cooking. For acorn meal cakes, I put the leaves on top of the rocks, cooking the meal. The cooked cakes can be easily separated from the rocks and the leaves peeled off. They can be eaten as-is or stored (if completely dry) at room temperature, sealed away from insects, for long-term use.

For acorn mush, I fill a wooden bowl with acorn meal and equal or greater portions of water. I sometimes add dried fruits. I place the hot stones in the bowl to cook the mush, sometimes wiping off the excess ash and char with a damp wad of fibers or cloth. It takes a few fist-sized hot stones to boil a large bowl of water. I remove the stones after wiping the adhering meal back into the bowl. I season the mush with pecan oil, salt, and sometimes maple syrup. It is delicious.

Quercus buckleyi Nixon & Dorr
Texas red oak
= Quercus rubra var. texana, Q. texana
Buckley oak, Nuttall oak, Texas oak, Spanish oak, spotted oak, rock oak, *chêne rouge*
Loc.: C, N, & SW TX; very common in Travis Co.
Form: tree, up to 140 ft. tall; perennial.
Food: no records.

Quercus fusiformis Small
Texas live oak
= Quercus virginiana var. fusiformis, Q. virginiana ssp. fusiformis
Escarpment live oak, plateau oak, Hill Country live oak, *encino prieto, encino molino, tesmoli*
Loc.: C, S, SW, & sparse E & N TX; very common in Travis Co.
Form: shrub, tree, up to 20-40 ft. tall; perennial.
Food: no records.

Quercus macrocarpa Michx.
Bur oak
Burr oak, overcup oak, savanna oak, mossy-cup oak, blue oak
Northern Cheyenne: *vóʔome-oóʔmeshe*, Hocąk: *piksigu*, Ojibwe: *mîtîgo' mîc* – "wooden tree"
Loc.: C & E TX; common ornamental in Travis Co.
Form: tree, up to 150 ft tall; perennial.
Food: Northern Cheyenne, Hocąk, Ojibwe.

Quercus marilandica (L.) Münchh.
Blackjack oak
Barren oak, black oak
Comanche: *ḓuhu:pᵛ / tuhuupi / tuhuupit* – "black wood"
Loc.: E, C, & S TX; uncommon in Travis Co.
Form: tree, up to 30-95 ft. tall; perennial.
Food: Comanche.
Notes – The Comanche ate these acorns raw or boiled.

Quercus muehlenbergii Engelm.
Chinquapin oak
= Quercus acuminata, Q. alexanderi, Q. prinoides
Chinkapin
Loc.: C, E, & W TX; common in Travis Co.
Form: tree; perennial.
Food: Cherokee.

Quercus sinuata Walter
Bastard oak

Durand oak
Loc.: all TX except W, far S, & far N; common in Travis Co.
Form: tree; perennial.
Food: no records.

Quercus stellata Wangenh.
Post oak

= Quercus minor, Q. obtusiloba
Iron oak, cross oak
Kiowa: *tdok-a-die-añ* – "oak" / *doʔgo't'ä'* / *do-ʔgo-tä* – "very hard wood"
Loc.: C, E, SE, & N TX; common in Travis Co.
Form: tree, 30-100 ft tall; perennial.
Food: Gosiute, Kiowa.
Notes – The wood of this tree is popular for smoking local barbecue.

Quercus virginiana Mill.
Live oak

= Quercus sempervirens, Q. virens
Coastal live oak, southeastern live oak, Virginia live oak, *chéne vert*
Loc.: C, SE, & E TX; common in Travis Co.
Form: tree, 40-80 ft. tall; perennial.
Food: Southeast Natives (eaten in large quantities).
Notes – It hybridizes with *Quercus fusiformis*, its closest related species, and the two are difficult to distinguish. The tips (apex) of its acorns are blunted as opposed to the more pointed ones of *Q. fusiformis*.

Additional notes:

Oak wood had many uses among Indigenous peoples in North America. It was used for fire-drills, cooking and smoking fuel, building material, digging sticks, utensils, bowls, mortars, tools and handles, basketry, rabbit sticks, clubs, lances, arrows, and bows.

Oak wood was used for bows by Florida Natives in 1540. These bows were reportedly capable of shooting completely through four layers of high-quality steel chainmail hauberks, or the entirely through a horse, lengthwise. Oak bows were also used by the Cherokee, Diné, Hopi, Tewa, and likely many other Indigenous peoples.

A decoction of the bark of *Q. fusiformis*, *Q. marilandica*, *Q. rubra*, *Q. stellata*, *Q. texana*, and *Q. virginiana* was used as a red dye or paint for basketry, buckskins, and more (Caddo, Choctaw, Houma, Forest Potawatomi). The leaves of *Q. marilandica*, *Q. nigra*, and *Q. stellata* were used for rolling tobacco for smoking (Comanche, Kiowa).

Quercus fusiformis. Base photo: tree overhanging excellent acorn gathering spot in dry creekbed in mid-October. Overlay, from top right: range map (GBIF); foliage and immature acorns in late September; harvested, mature acorns in late October.

Top half: *Quercus macrocarpa* foliage; acorn in late September; range map (GBIF).
Bottom half: *Quercus muehlenbergii* foliage; harvested acorns in mid-October; range map (GBIF).

JUGLANDACEAE – Walnut Family

Carya illinoinensis (Wangenh.) K.Koch
Pecan

= Carya oliviformis, C. pecan, Hicoria pecan
Comanche: *nakutβai / nakkʉtab?I / nakkʉtab?i* ~ "crack it with your teeth," Kiowa: *on-ku-a / do'nä'i / don-ai* – "fat (or oily) tree fruit," Osage: *wataθtθta*
Loc.: all TX, sparse N, S, & W; very common in Travis Co.
Form: tree, up to 70-160 ft. tall; perennial.
Flowers: Mar-May (yellow).
Notes – This is the state tree of Texas.

History

Seeds – Pecan nuts were highly prized by the Asinai, Caddo, Cenis, Comanche, Karankawa, Mexican Kickapoo, Kiowa, Nabedache, Natchitoches, Nazones, and Osage, other Natives, folk in Northern Mexico, and Texians. They were eaten in large quantities and were stored for the winter (Comanche, Osage). Pecans were traded by Natives and among Texians; the pecan trade was an early industry in Central Texas.

Carya (hickory) species in general were an important food source for Indigenous peoples of the Eastern US, with nuts being eaten and stored in great quantities throughout the Southeast, according to early Spanish expeditions. Hickory nuts were also eaten by the Cherokee, Choctaw, Dakota, Haudeonsaunee, Menominee, Meskwaki, Omaha, Ojibwe, Pawnee, Ponca, Forest Potawatomi, Quapaw, and Winnebago, as well as by Indigenous peoples in all regions of North America where *Carya* species occur. Other species eaten include *C. alba*, *C. cordiformis*, *C. laciniosa*, and *C. ovata*.

Natives in south-central Texas in 1709 used "nuts" for food "for the greater part of the year." Of the various kinds of nuts eaten, "all of them are more tasty and palatable than those of Castile, though they are longer and thinner." This is likely a reference to *Carya illinoinensis*, a common wild species in the area, but may have also included *Carya texana*. Large quantities of the whole nuts were stored in underground caches. The people were "very skillful at shelling them, taking the kernels out whole." The shelled nuts were stored in "small sacks made of leather for the purpose," and were sometimes also kept threaded on long strings.

Hickory nuts were eaten plain, in soups, with honey, mixed with cornmeal for bread, as mush, with beans or berries, or processed into oil. Nut milk was also made (Choctaw, Haudenosaunee).

Hickory nuts were commonly made into oil (Cherokee, Choctaw, Haudenosaunee, Southeast Natives). The general method of oil extraction was to place the nuts in small, shallow cavities pecked into a large flat anvil stone, and strike them with a hammerstone. Sometimes this was done with multiple anvil cavities so that up to five or more nuts could be cracked in a single blow. Without separating the nut meat from the shells, the cracked nuts were boiled and the oil that rose to the top was skimmed off and stored in earthenware jars. This oil was used for cooking, especially in corn cakes. It was also used as a medium for paints, for preserving leather, for skin and hair care, and for preserving and polishing various implements such as bows. One hundred pounds of nuts yielded 154 ounces of oil.

The Haudenosaunee ground the nuts and boiled the meal. They skimmed off the oil, boiled it separately, and salted it for use in bread, potatoes, squash, and other foods. The leftover nut meat was often seasoned and mixed with mashed potatoes or corn pudding. In another account, the Haudenosaunee cracked the nuts on special small stone boulders or slabs. Each stone had a small depression pecked into it to hold the nut in place as it was cracked with another stone. They also ground the whole nuts, shells and all, in a large wood mortar. The meal was mixed with water, allowing the shells to sink to the bottom. The upper portion of milky water (nut milk) was drunk, being called "*pawcohiccora*" or "*patahikarea*" (Lenape), possibly the origin of the word "hickory." The Seneca mixed dried hickory nut meat meal with pulverized dried bear or deer meat. This mixture was boiled and used as baby food.

The Choctaw extracted oil by parching the nuts and grinding them into a fine meal. This meal was boiled for one to one and a half hours, after which the oil was strained off and the remaining solids discarded. The oil was used for cooking.

The Cherokee ground the nuts, shells and all, in a large mortar. The meal was placed in a container with water, and the oil that rose to the top was skimmed off. Heat may have been applied, but this was not mentioned. This hickory oil, called "*canuchi*," was used as cooking oil and to make cornbread.

Leaves – Pecan leaf tea was favored by the Mexican Kickapoo.

Notes

Character – *C. illinoinensis* is the only commonly occurring *Carya* species in Travis County (*C. texana* is also present, but very rare) and has the best nuts in the genus, with thin shells and plenty of nut meat. Other hickory nuts are harder to crack and extract the nut meat.

Pecan trees look very similar to walnut trees (*Juglans major* or *J.*

nigra), especially their foliage, but can be distinguished by smell. The scent of crushed pecan leaves is distinctly fragrant, though this may only be judged by personal experience. *Juglans* species typically have more leaflets than *Carya*, although there is some overlap in numbers. Pecan trees have 9-17 leaflets, while the three walnut species in Austin have 7-25 (*J. microcarpa* has the most, *J. major* the fewest).

The best characteristic for distinguishing the non-reproductive parts of *Carya* and *Juglans* can be seen by splitting the twigs lengthwise. *Carya* twig pith is solid and homogeneous, whereas *Juglans* pith is chambered, exhibiting lines of separation. The fruits can be easily distinguished: pecan hulls are green, oblong, and divided into four segments, whereas walnut hulls are green, globular, and unsegmented.

Season – Peak pecan season in Austin is in the fall, especially around mid-October. Pecans can be found from late summer to early winter.

Nutrition – Raw pecans have 10% protein, 73% fat, 13% carbohydrates, and 6% fiber.[70] They contain potassium (360 mg/100 g), magnesium (103 mg/100 g), zinc (4 mg/100 g), manganese (2 mg/100 g), iron (2 mg/100 g), and copper (1 mg/100 g).[70] They are excellent sources of copper, manganese, zinc, and magnesium, and moderate sources of iron and potassium.

Pecan oil has good oxidative stability, and pecan nut consumption can decrease total cholesterol.[63] Pecans are rich in healthy fats such as linoleic and oleic acids (polyunsaturated and monounsaturated fats), polyphenolic compounds such as epigallocatechin gallate (an antioxidant found in green tea), and tocopherols (vitamin E, 12 mg/100 g), making them beneficial to cardiovascular health, obesity management, oxidative stress reduction, and combating metabolic disease.[10] These figures are from commercial pecans. Wild pecans are slightly smaller on average and may contain somewhat lower macronutrient levels, but their mineral, vitamin, antioxidant, and lipid profiles are essentially similar.

Practice – Similar to acorns, pecans are difficult to gather from the floor of an intact forest. Fortunately, like oaks, pecans are popular landscaping plants in Austin, so the grounds beneath them are often cleared, making them easy to gather. I also like to harvest pecans directly from the trees. As they ripen, the green hulls begin to split open, revealing pecans that have yet to drop. During this brief period, around early October, it is easy to collect nuts directly from the branches. I often bring a long hooked stick that I use to pull down branches full of nuts.

I gather many pecans while they are still in their hulls, even before the hulls split open, but many of these turn out to be "duds," pecans that the tree aborted during development and are essentially empty shells. There are two types of duds. One is easily identifiable, with a smaller hull that is often brown or discolored. The other type has a hull similar to a ripe, healthy pecan but never splits open. These can be detected by the lack of hull-splitting and a slightly lighter weight. I sometimes collect these latter duds accidentally and only identify them once they have been setting for a time and never split open. When hulled, they have a full-sized shell that looks like a normal pecan but weighs less and is hollow, containing no nut meat, or only a small amount of shriveled, dry, and bitter tissue.

I slip off any green hulls (which can be used for a dark brown dye) then wash the pecans by putting them in a tub of water and scrubbing them with wads of grass. I dry them completely before long-term storage in the shells. To eat them, I crack the shells (with the point upward) with a rock on an anvil stone and tease out the nut meat by hand or with a long antler or bone pick. In the field, I crack pecans with my teeth (not recommended) or by pressing two together in my palm until they break. I eat the nut meat plain or in various dishes, desserts, or smoothies.

I sort out the smallest grade of pecans from the batch to make oil. I crush these with my large mortar and pestle, reducing them to a coarse meal. I boil this meal for a while, causing the oils to separate. I let the decoction cool, then strain it through cloth. I put cotton cloth in a colander, then pour the decoction and hot meal into it. I twist up the sides of the cloth to seal it, then squeeze the bundle to force out the rest of the water and oil from the meal. Sometimes, I put the strained and pressed pecan meal back into a pot of water and re-boil it to extract the maximum oil, but the first batch will extract most of it. The oil can be skimmed off the top of the strained liquid, but some water will be included. This extra water can be cooked off at a low simmer. Another way to separate the oil from the water is to freeze it. Once the water freezes, the oil can be cleanly poured off the ice. This oil (if water-free) can last almost a year without spoiling, especially if refrigerated.

I use pecan oil in place of any cooking oil or fat. It is excellent as a replacement for butter in recipes. It also makes a high-quality wood sealant. It can be used to moisturize skin, as a carrier oil for an extract of American beautyberry (*Callicarpa american*) for mosquito repellent, or for countless other purposes.

Carya illinoinensis. Base photo: large tree in late September. Overlay, clockwise from top right: range map (GBIF); harvested pecans in mid-November; hull opening on tree to release pecan in mid-October; unopened cluster of fruits (note that the lowest one is a "dud") in mid-August; foliage (strong floral smell when crushed); pecan oil.

Juglans spp.
Walnut

Comanche: *mubitai*, Tewa: <u>ko</u> *'ǫnto* – "buffalo nut"
Loc.: W, C, E, & sparse S & N TX; common in Travis Co. (3 spp.); 4 spp. in TX.
Form: shrub, tree; perennial.

History

Seeds – Walnuts were eaten by the Chiricahua and Mescalero Apache, Cenis, Cherokee, Comanche, Dakota, Diné, Haudenosaunee, Hocąk, Mexican Kickapoo, Kiowa, Menominee, Meskwaki, Ojibwe, Omaha, Osage, Pawnee, Kashaya Pomo, Ponca, Forest Potawatomi, Quapaw, Tewa, Tonkawa, Winnebago, Florida Natives in 1540, Southeast Natives, other Natives, Texas Natives 5,000 years ago, and Texas Natives in the Archaic. Species eaten include *J. cinerea*, *J. hindsii*, *J. major*, *J. microcarpa*, and *J. nigra*, which are all common native species; the final native, *J. californica*, was likely eaten as well.

Walnuts were generally grouped with, and eaten like, hickory nuts by Indigenous North Americans, especially in the Southeast, where both commonly occur. Walnuts were an important food source for Indigenous peoples in the Eastern US, being eaten stored in great quantities by Natives throughout the Southeast, according to early Spanish expeditions. They were often stored for winter use (Comanche, Meskwaki, Kashaya Pomo, Forest Potawatomi). The trees were planted, transplanted, and cultivated by Southeast Natives for their nuts.

Walnuts were eaten plain, as soup, with honey, as mush, or as "bread" (Apache, Comanche, Dakota, Haudenosaunee, Kiowa, Omaha, Pawnee, Ponca). The green walnut hulls were removed by the Comanche, and the nut meat was extracted, dried and stored. The shells were cracked with a hammerstone, and the nut meat was extracted with a five-inch-long bone pick.

Whole walnuts, including the nut meat, hulls, and shells, were mashed into a fine meal and covered with water, then boiled (Western Apache). When filtered, the liquid was white, tasted like milk, and was very nutritious. The nut meat was ground into a meal by the Chiricahua and Mescalero Apache and often eaten with cooked agave hearts. *J. major* nut meat was sometimes boiled by the Chiricahua.

Walnuts were ground into meal and mixed with cornmeal for bread, soup, or mush, with beans or berries also added (Haudenosaunee). To make oil, the Haudenosaunee ground up the walnuts, boiled the meal slowly, skimmed off the oil and boiled it separately, then salted the oil for use in bread, potatoes, squash, and other foods. The leftover nut meat was often seasoned and mixed with mashed potatoes.

Walnut oil from *J. nigra* was commonly made by Southeast Natives. The general method of oil extraction was to place the nuts in small shallow cavities pecked into a large flat anvil stone, and strike them with a hammerstone. Sometimes, this was done with multiple anvil cavities and nuts so that up to five or more nuts could be cracked in a single blow. Without separating the nut meat from the shells, the cracked nuts were boiled and the oil that rose to the top was skimmed off and stored in earthenware jars. This oil was used for cooking, especially for corn cakes. It was also used for a paint medium, preserving leather, skin and hair care, and preserving and polishing various implements such as bows. One hundred pounds of nuts yielded 80 ounces of oil, more than half the volume of oil made with an equal weight of hickory nuts.

Notes

Character – *Juglans major* and *J. nigra* are both in Austin, and look very similar. The fruits of *J. major* are smaller (0.8-1.4 inches) than those of *J. nigra* (1.4-3.1 inches) and the green hulls of *J. major* have a smoother skin, whereas *J. nigra* hulls have a coarse texture. *J. major* has microscopic hairs on its leaflets, whereas those of *J. nigra* are mostly smooth. *J. major* has 9-15 leaflets, whereas *J. nigra* has 9-23 leaflets.

For the purpose of food, *J. nigra* is superior because its nuts are larger, but all three species are good food sources. *J. microcarpa* is less usable because it has the smallest nuts of the three, but these are still perfectly edible. As the name implies, they are like miniature versions of the other walnuts. *J. microcarpa* nuts are especially suited to making walnut oil, as it is less efficient to manually extract their nut meat compared to pounding them whole to boil for oil.

As an aside, similar to a pattern also seen in *Rhus trilobata* and *R. aromatica*, Central Texas is the only place where two widespread Southwest and Southeast species co-occur, namely *J. major* and *J. nigra*. Furthermore, it also has *J. microcarpa*, making it, apart from Comanche Co., OK, the only North American region with three native walnut species.

Season – Walnuts start reaching full size in midsummer, when they can be gathered by pulling them off the tree. Peak nut maturity is from late summer to early fall, when they start falling and can be gathered from the ground. They do not spoil quickly, so can be gathered well into the winter.

Nutrition – *J. nigra* walnuts are richer in vitamin C and potassium than the commercial walnut (*J. regia*) and rich in leucine and phosphorus.[7] J.

regia walnuts have 15% protein, 70% fat, 11% carbohydrate, and are rich in polyunsaturated and monounsaturated fats. They are rich in omega-6 and omega-3 (ALA) fats and vitamin E.[5,6] They are good sources (~9-34% DV) of potassium (424 mg/100 g), magnesium (142 mg/100 g), zinc (3 mg/100 g), and iron (2 mg/100 g), and excellent sources (~111-130% DV) of copper (1 mg/100 g), and manganese (3 mg/100 g).[7] *J. nigra* kernels have similar or higher levels of all of these minerals.[7]

Practice – With such an abundance of pecans, I do not feel the need to forage walnuts for regular subsistence, but they make an excellent food storage and backup nutrition source. I gather all three species. They basically do not spoil from storage. I have stored a big pile of them for a year outdoors and they are perfectly edible once cracked open, despite not even removing the green hulls, letting them rot and dry into detritus.

I either crack them and pick out the nut meat to eat or crush them whole and boil them into an oil, following the same process I describe for pecans. The green hulls make an excellent dark brown dye.

Juglans major (Torr.) A.Heller
Arizona black walnut

= Juglans elaeopyren, J. microcarpa ssp. major, J. rupestris var. major
Texas black walnut, river walnut, mountain walnut, *nogal silvestre*
Apache: *hałtsede* – "that which one breaks," Diné: *xa'ałtshétiih* – "that which is cracked"
Loc.: C, W, & sparse N TX; not uncommon in Travis Co.
Form: tree, 36-48 ft. tall; perennial.
Food: Chiricahua and Mescalero Apache, Diné.

Juglans microcarpa Berl.
Little walnut

Texas black walnut, Texas walnut, river walnut, *nogalito, nogalillo, namboca*
Loc.: C, W, & sparse N TX; common in Travis Co.
Form: shrub, tree, up to 20-30 ft. tall; perennial.
Food: Mexican Kickapoo, Texas Natives 5,000 years ago, Texas Natives in the Archaic.

Juglans nigra L.
Eastern black walnut

Black walnut, American black walnut, *l'arbre à noisette*
Cayuga: *nyugwagwi′noni'*, Cherokee: *sedí / setí*, Comanche: *muβitai / tuhmuβitai*, Dakota: *hma*, Teton Dakota: *cha^n-sapa* – "black wood," Haudenosaunee: *djonyot'gwak*, Hocąk: *cakhu*, Kiowa: *poho'n-ä, po-ho-na / poñ-hoñ-á-daw / poñ-hoñ-ai-gaw*, Meskwaki: *pakan / pûka'nak / makwe'paka'nanį^h*, Omaha-Ponca: *tdage*, Onondaga: *deyutsu`'gwagwi''noni'* – "round nut," Osage: *tage*, Pawnee: *sahtaku*, Winnebago: *chak*
Loc.: C & E TX; common in Travis Co.
Form: tree, up to 50-150 ft. tall; perennial.
Food: Cherokee, Comanche, Dakota, Haudenosaunee, Hocąk, Kiowa, Meskwaki, Omaha, Osage, Pawnee, Ponca, Tonkawa, Winnebago, Southeast Natives, other Natives.

Juglans spp. Base photo: *J. microcarpa*. Overlay, clockwise from top left: *J. microcarpa* fruits in early July; *J. nigra* fruit in late June; *J. nigra* nut meat (aged 1 year); *J. nigra* tree; *J. nigra* flowers in late April. Range map (*Juglans* spp.): GBIF.

LAURACEAE – Laurel Family

Lindera benzoin (L.) Blume
Northern spicebush

= Laurus benzoin
Cherokee: *nɔʔdaˡlí / noʔstaᵈgí / nɔˑtaˡsí*, Creek: *kəpəpakə*
Loc.: C & E TX; uncommon in Travis Co.; only *Lindera* sp. in TX.
Form: shrub, tree, up to 15-20 ft. tall; perennial.

History

Twigs – A bunch of twigs, as big as a fist, was boiled for a short time and the decoction was drunk by the Cherokee as a beverage.

Notes

Character – The leaves, twigs, and foliage of this plant are exceptionally aromatic, with a pleasant, medicinal, spice-like smell. The fruits are also a favored spice-like food among foragers.

Season – The leaves are deciduous, but the twigs can be used for tea year-round. The fruits ripen from late summer to fall.

Nutrition – The potent aroma of the foliage is from its essential oils.[66] The leaves, twigs, and fruits vary significantly in essential oil composition, with dominant compounds having little overlap between tissues.[66]

Practice – The leaves and twigs make a flavorful tea.

Lindera benzoin foliage. Range map: GBIF.

MORACEAE – Mulberry Family

Morus distribution. Source: GBIF.

Morus spp.
Mulberry

Comanche: *etɨɨai* (*etɨɨ* – "bow")
Loc.: all TX; common in Travis Co. (3 spp.); 4 spp. in TX (2 introduced; *M. alba* & *M. nigra*).
Form: shrub, tree; perennial.

History

Fruits – Mulberry fruits were eaten by the Chiracahua and Mescalero Apache, Asinai, Caddo, Comanche, Haudenosaunee, Quapaw, Tohono O'odham, Natives of Illinois prairies, Natives in south-central Texas in 1709, Florida and Kansas Natives in 1540, Natives in South Carolina in 1567 (likely Creek, Catawba, Cherokee, Yamasee, Wateree, or Santee), and Natives in Missouri, Kansas, and Oklahoma. Species eaten include *Morus alba*, *M. microphylla*, and *M. rubra*.

May was "mulberry month" or "*hàsh bihi*" in Choctaw. They were found in great abundance along the Wateree River, and the area was called "*Guiomae*," in the 1500s, perhaps derived from the Muskogee "*ki-o-mays*," meaning "a place where there are mulberries" (Hudson 1990). Mulberry fruits were also eaten by members of the La Salle expedition to Texas. Indigenous peoples traveled long distances to gather the fruit.

Notes

Character – Similar to its unique abundance in *Rhus* and *Juglans*, Texas, especially Central Texas, is remarkable for having three *Morus* species, unlike every other area of North America (except Comanche Co. and Murray Co., OK), which typically has one native species (*M. rubra* or *M. microphylla*) plus the introduced, and somewhat invasive, *M. alba*, especially in urban areas. All *Morus* species have aggregate fruits.

All three *Morus* species are relatively common in Austin, though as a large urban area, *M. alba* is probably the most abundant. *M. microphylla* prefers Edwards Plateau habitats with drier limestone soils, whereas *M. rubra* prefers Blackland Prairie habitats with moist, clay soils.

M. rubra attains a much larger size, whereas *M. microphylla* is usually a shrub or a small tree. The foliage also distinguishes the two: *M. microphylla* has much smaller, more lobed leaves, whereas *M. rubra* leaves are very large, usually larger than a hand, and are often not lobed, although most trees have both lobed and unlobed leaves. *M. alba* reaches a smaller maximum size than *M. rubra*, but is generally larger than *M. microphylla*. *M. alba* can be easily distinguished from *M. rubra* by its glossy leaves, compared to the matte leaves of *M. rubra*. *M. alba* leaves are typically more lobed than *M. rubra*. *M. alba* and *M. rubra* occasionally hybridize. These hybrids can be difficult to distinguish, exhibiting intermediate traits such as large, slightly glossy, moderately lobed, or sometimes entire leaves.

Season – In Austin, mulberries can be found ripening in early to mid-spring, peaking in late spring, and can usually no longer be found by midsummer.

Nutrition – The fruits of *M. alba* and four other wild *Morus* species in Pakistan have an average of 10% protein, 11% fat, 75% carbohydrate, and 5% fiber, with 432 kcal/100 g.[1] They are rich in potassium, magnesium, iron, sodium, and zinc and good sources of phenolics, flavonoids, and vitamin C.[1]

Practice – Mulberries are the wild fruit I have been foraging the longest, since early childhood. They are very easy to identify, common, and delicious. I simply gather them by hand. I prefer to eat them fresh and whole, but they can also be made into preserves, or canned or dried for storage.

Morus alba L.
White mulberry

Loc.: all TX except far W; common in Travis Co.; introduced.
Form: shrub, tree; perennial.
Food: Asinai.

Morus microphylla Buckley
Texas mulberry

= Morus confinis, M. crataegifolia, M. grisea, M. radulina
Littleleaf mulberry, mountain mulberry, Mexican mulberry, dwarf mulberry
Apache: *tsełkane*, Tohono O'odham: *kohi*
Loc.: all TX (more in C, W, & S); not uncommon in Travis Co.
Form: shrub, tree, up to 25 ft. tall; perennial.
Flowers: Mar-May (red, green).
Food: Chiricahua and Mescalero Apache, Tohono O'odham.

Morus rubra L.
Red mulberry

Comanche: *etehub^v* / *etɨhuupl* – "bow wood" / *sɔhɔβoko* / *soho boʔko*, Haudenosaunee: *odji'nowǒ^n'wadisiya djoyesshăyes*, Meskwaki: *wasao'kûs*, Mohawk: *deyuderaha'kdǫ*
Loc.: all TX except NW; common in Travis Co.
Form: shrub, tree, up to 65 ft. tall; perennial.
Food: Asinai, Comanche, Haudenosaunee, Natives in South Carolina in 1567 (likely Creek, Catawba, Cherokee, Yamasee, Wateree, or Santee), Natives in Missouri, Kansas, and Oklahoma. Note that this is the only native mulberry species in Eastern North America, and was likely eaten by Indigenous peoples throughout its range.

Additional notes:

Mulberry wood, especially, and perhaps only, from *M. microphylla*, was preferred for making bows among the Chiricahua and Western Apache, Comanche, and Tohono O'odham. The Comanche considered it second only to Osage orange (*Maclura pomifera*), which is widely regarded as the best bow wood in North America and is also in the mulberry family.

The Apache cut a straight, knot-free trunk or limb, trimmed it into a stave while still green, and then hung it up to dry for several weeks. They sometimes recurved the ends after finishing the tillering.

The Comanche selected straight, young trees, and peeled off the bark before drying the wood. After trimming the staves to shape, they were sanded smooth with sandstone. The wood was greased and heated over a fire to shape it, sometimes being braced in a forked branch to bend it. Selfbows (straight until strung) were preferred for war, while recurved bows (ends curved away from shooter) were preferred for hunting. They measured about three and a half feet long and were notched at the ends to receive the bowstring. Bear intestines were used for the string, as they were resistant to wet weather.

The Tohono O'odham cut the wood in the rainy season (August in their area), selecting a straight shoot about five feet long and two inches in diameter. After removing the bark, it was trimmed to shape, sanded with a stone, and notched. The stave was shaped by laying it in hot ashes and using stones placed on and around it to form the proper curve. The ends were bent up at a gradual angle, but the central part was not bent at all since that would weaken it. The ends were then tied in position with agave fiber cord, and set to thoroughly dry.

I have made a selfbow from a green *M. alba* branch with only stone tools. It is very strong and elastic, and capable of sufficient penetration for hunting despite its small size (3.5 ft. long).

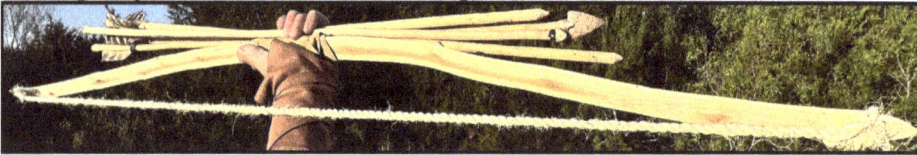

Bow made from *M. alba* branch with agave fiber bowstring and roughleaf dogwood arrows with stone and antler points. All materials were gathered in Austin and only stone tools were used.

Base photo: *Morus microphylla*. Overlay, clockwise from top left: *M. microphylla* range map (GBIF); *M. alba* foliage, range map (GBIF), and ripe fruit in late April. *M. microphylla* bark, unripe fruit in late March, and foliage. Note the leaves of *M. alba* are glossy, thick, and larger than those of *M. microphylla*.

Morus rubra. Base photo: full-size tree. Overlay, clockwise from top left: bark of young tree; range map (GBIF); young foliage & flowers in late March; leaves; and leaves of a hybrid *M. rubra* x *alba* (note the size and shape are more characteristic of *M. rubra*, but the glossiness is more characteristic of *M. alba*). All *Morus* spp. leaves can be deeply lobed (not shown).

OLEACAEAE – Olive Family

Forestiera pubescens Nutt.
Stretchberry

= Forestiera neomexicana, F. sphaerocarpa
Elbowbush, desert olive, downy forestiera, wild privet, spring goldenglow, ironwood
Apache: *iɣentłidzi* – "hard seed," Diné: *k'įicjíníh* / *mą'įitqa'* – "coyote food" / *mai'ii dąą*', Hopi: *ɜtɜ'svi*, Jemez: *uiņúʃ*, Kawaiisu: *toṁbovi*
Loc.: C, W, & N TX; very common in Travis Co.
Form: shrub, up to 15 ft. tall.; perennial.
Flowers: Jan-June (yellow, green).

History

Water – These shrubs were considered by the Isleta Pueblo to be water indicators; wells dug where they grew consistently produced water.
Fruits – Stretchberries were eaten fresh by the Chiricahua and Mescalero Apache. They were known by the Kawaiisu to be "eaten by coyote and bear," though they did not consume them themselves.

Notes

Character – The name "elbowbush" refers to the tendency of this plant's branches to form roughly 90-degree angles with the main stem, a useful diagnostic feature of this otherwise somewhat unremarkable common shrub. Its branching is opposite, despite the tendency for it to superficially appear alternate, with branches and petioles often slightly offset.

The specific epithet "*pubescens*" refers to the slight fuzziness characteristic of the leaves. *Ligustrum* species appear similar but have glossy leaves.

The flowers are polygamo-dioecious, meaning most plants have male and female flowers on separate individuals but some individuals have perfect (male and female) flowers, and others may even have a mix of male, female, and perfect flowers.

Season – The ripe fruit availability peaks in early May and they can be found from mid-April to late May.
Nutrition – The fruits of *Forestiera angustifolia* contain 7% protein and were rich in potassium, with 1600 mg/100 g.[26]
Practice – These fruits are not well-known as edible, but they have a unique and appealing taste. They are sweet and juicy with floral notes and a faint astringent aftertaste. The first time I tried them, I was unsure if I liked them, but after eating them a few times, I began to appreciate them. I gather large amounts in my foraging net with the "strike and catch" method. I eat them whole and raw, cooked into a sauce, boiled to make a beverage, or dry them for storage. I eat the dried ones whole, in smoothies, as tea, or rehydrated.

Forestiera pubescens. Base photo: fruiting plant in early May. Overlay, clockwise from top left: range map (GBIF); male & female flowers in early March; ripe fruits in early May; harvested fruits in mid-May; foliage (note 90° branches).

Fraxinus pennsylvanica Marshall
Green ash

= Fraxinus campestris, F. darlingtonia, F. lanceolata, F. smallii
Downy ash, red ash, swamp ash, river ash, water ash, Darlington ash
Northern Cheyenne: *moʔtóʔe*, Dakota: *psehti^n*, Ojibwe: *a' gîma'k* – "snowshoe wood," Omaha-Ponca: *tashnánga-hi*, Pawnee: *kiditako*, Forest Potawatomi: *êmkwansûk* – "spoon wood," Winnebago: *rak*
Loc.: E, NE, SE, & C TX; very common in Travis Co.
Form: tree, up to 50-75 ft. tall; perennial.

History

Inner bark – The inner bark was scraped down in long, fluffy layers and cooked by the Ojibwe. It was said to taste like eggs, and the Ojibwe name for the food, *sagîma' kwûn* (*wûn* – "eggs"), roughly means "ash tree eggs."

Notes

Character – The "keys" or samaras (winged seed pods) are edible raw or cooked when young and soft, and are sometimes pickled by modern foragers.

Fraxinus albicans (Texas ash) is common in the area and can be distinguished by its rounded leaflets and smaller maximum size. *F. berlandieri* also occurs in the area and tends to have fewer and smaller leaflets, but overlaps in characteristics with *F. pennsylvanica* and can therefore be difficult to distinguish.

Practice – I have tried scraping off and eating the inner bark a dozen or more times and it was unpleasantly bitter each time. I still ate it, but it did not seem palatable or nutritious. I may not be harvesting it in the proper season or habitat, or perhaps I am not processing it correctly. Boiling it, especially with a water change, improves the taste. The best season for harvesting is likely late winter or early spring, when the sap is rising.

The young samaras of green ash are a nice snack, and are produced prolifically. The best stage for eating them is just before they reach full size, when they are still soft and flexible. Once mature, they begin to dry out and stiffen, becoming fibrous and inedible by midsummer.

Additional notes:

The wood of this species was used for basketry (Ojibwe, Forest Potawatomi) as well as for bows and arrows (Northern Cheyenne, Dakota, Omaha, Pawnee, Ponca, Winnebago). The sapwood was split, pounded, and peeled to obtain basketry splints.

Fraxinus pennsylvanica. Base photo: fruiting plant in early April. Overlay: young samaras in early April; range map (GBIF).

PHYTOLACCACEAE – Pokeweed Family

Phytolacca americana L.
Pokeweed

= Phytolacca decandra
Poke, poke berry, inkberry, redweed
Cherokee: *dzayitagá* / *dza¹dagá* / *dza·yadehí*, Haudenosaunee: *o"sheä onĕ"'ta'* – "crimson leaves,"
Kiowa: (name translates to) "pink flower plant," Osage: *gðebe moŋkoŋ* – "vomit medicine"
Loc.: E, C, & sparse W TX; common in Travis Co.; only *Phytolacca* sp. in TX.
Form: herb, shrub, up to 5-20 ft. tall; perennial.
Flowers: Jan-Dec (white, pink, purple).

History

Foliage – The young, tender foliage was gathered in the spring, cooked, and eaten by the Cherokee, Kiowa, and Haudenosaunee.

Notes

WARNING – All parts of pokeweed are highly poisonous. It has historically been eaten by folk in the Southeast US, but only after proper cooking. Recommended methods for cooking pokeweed vary. Unless you are certain of a reliable method, I do not recommend experimenting with eating pokeweed.

Practice – I gather the young spring foliage, both leaves and stems, thoroughly boil them, discard the water, and eat the cooked foliage. They have a neutral taste, with a tender, moist texture. With some salt, they make a palatable green vegetable. They are commonly combined with egg dishes among Southeast folk.

Additional notes:

The fruits and seeds were used by the Cherokee and Kiowa to treat rheumatism. The fruits were simply eaten or crushed, sugar was added, and allowed to ferment; a tablespoon of this "poke berry wine" was taken at a time. A decoction of the roots was applied to eczema (Cherokee). The roots were used by the Osage as an emetic and laxative, hence the Osage name for the plant. Every spring, the Osage men gathered to chew the root. This was perhaps a ritual cleansing similar to that performed by many Texas tribes in their first fruits ceremony.

The dark red fruits were used as a dye by the Caddo, Comanche, Kiowa, and Ojibwe, especially for dyeing basketry and feathers. Feathers were dyed by the Comanche by mashing the fruits in an animal stomach, putting the feathers in, and burying the mass for four days. The dried fruits were sometimes used as necklace beads (Kiowa).

Phytolacca americana. Base photo: young plant in late March. Overlay, clockwise from top right: flower in late April; fruits in mid-August; range map (GBIF).

RHAMNACEAE – Buckthorn Family

Sarcomphalus obtusifolius (Hook. ex Torr. & A.Gray) Hauenschild
Lotebush

= Rhamnus obtusifolia, Ziziphus lycioides, Z. obtusifolia
White crucillo, *capulín*, *bachata*
Tohono O'odham: *u:s tcui'tpa't*, Seri: *xica imám coopol* – "black-fruited things" / *haaca*
Loc.: all TX except far E & far N; not uncommon in Travis Co.; only *Sarcomphalus* sp. in TX.
Form: shrub or small tree, up to 8 ft. tall; perennial.
Flowers: Mar-Apr (yellow, green).

History

Fruits – Lotebush fruits were eaten by the Mexican Kickapoo, Seri, and Tohono O'odham. They were eaten fresh by the Seri and were often raided from pack rat (*Neotoma*) nests. The Tohono O'odham boiled the fruits into a syrup to eat. This was prepared the same as they did prickly pear fruit syrup. Lotebush fruits were put in a basket and mashed into a pulp with a stick, then the juice was squeezed out, strained twice, then boiled and strained again to make the syrup. The pulp juice was also sometimes fermented (Tohono O'odham).

Notes

Character – These fruits are relatively unknown as edibles, despite the wide range of the species throughout Texas, the Southwest, and Northern Mexico. Each fruit is composed roughly of half juicy flesh and half seed. *Ziziphus jujuba* is the commercial jujube (Chinese date) and several other *Ziziphus* species are commercially cultivated.[9] Lotebush is closely related to this genus, in which it was formerly classified.

Season – The peak availability of ripe lotebush fruits in Austin is the first half of May, but they can be found from April to June.

Nutrition – *Ziziphus mauritiana* fruits are rich in vitamins A and C.[9] *Z. jujube* fruits contain about 6% protein, 83% carbohydrate, 8% fiber, potassium (291 mg/100 g), and vitamin C (273 mg/100 g), with high antioxidant capacity.[48] This is over five times the vitamin C content of oranges, suggesting that lotebush fruits are also likely exceptional sources of vitamin C.

Practice – The fruits are sweet, fleshy, fruity, semi-juicy, and date-like. I eat them fresh and raw, straight from the bush, or dry them for storage. They can also be cooked into a pulp, strained, and canned, made into preserves, or spread on a sheet and dried for fruit leather.

Sarcomphalus obtusifolius. Base photo: fruiting plant in mid-May. Overlay, clockwise from top right: range map (GBIF); flowers in late May; leaf; ripe fruits in mid-May.

ROSACEAE – Rose Family

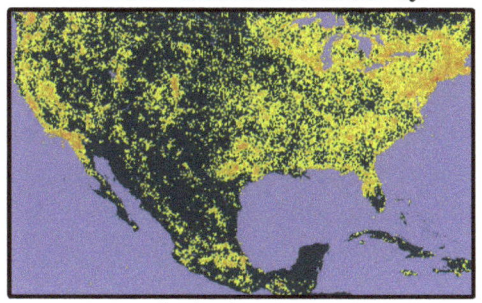

Prunus distribution. Source: GBIF.
Prunus spp.
Plum

Cherry, peach, apricot, almond
Comanche: *yusəkə* (early plum) / *parawasəkə* (late summer plum) / *kusiəkə* (fall plum) / *sʉkʉʔi* / *pibiarona* / *natsomʉ* (dried plums), Haudenosaunee: *oyă'gane gowa*
Loc.: all TX; common in Travis Co. (7 spp.); 18 spp. in TX (3 introduced).
Form: shrub, tree; perennial.

History

Fruits – Wild plums and cherries were eaten by the Chiricahua, Jicarilla, and Mescalero Apache, Arapaho, Asinai, Blackfeet, Caddo, Cahuilla, Calpella, Cherokee, Northern Cheyenne, Choctaw, Comanche, Concow, Cree, Dakota, Diné, Flathead, Gosiute, Haudenosaunee, Hidatsa, Hocąk, Isleta, Kawaiisu, Kiowa, Klamath, Lakota, Little Lake, Northern Maidu, Menominee, Meskwaki, Natchitoches, Oglala, Ojibwe, Omaha, Pawnee, Pomo, Ponca, Forest Potawatomi, Nomlaki, Quapaw, Shasta, Sioux, Tewa, Northern Ute, Wailaki, Winnebago, Yokia, Yuki, Yurok, Indigenous people in Tenochtitlan around 1519, Natives in Florida, South Carolina, Arkansas, and Kansas in 1540, Carolina Natives in 1700, Natives in the Pacific Northwest, Southeast, West, and Texas. Species eaten include *Prunus americana, P. andersonii, P. angustifolia, P. ilicifolia, P. mexicana, P. nigra, P. pensylvanica, P. persica, P. pumila, P. serotina, P. subcordata, P. susquehanae,* and *P. virginiana.*

Prunus fruits were some of the most important fleshy fruits in the historical diet of American Indians. From the earliest European contact in the 1500s, Southeast Natives were eating, drying, and storing *Prunus* fruits in great quantities and cultivating them in orchards. These fruits were highly valued food across North America and were gathered in large quantities to be eaten fresh, cooked, or dried for storage. Produce from various *Prunus* species was sold in markets by Indigenous people in Tenochtitlan around 1519. During the Texas Spanish Mission period, and possibly earlier, *Prunus* species, possibly *P. angustifolia* or *P. mexicana*, were being cultivated by the Asinai and Caddo in the Texas. *P. angustifolia* was commonly cultivated by Natives in the Southeast. The ripe fruits of *P. virginiana* were used by the Dakota to define a lunar month, called *Ca^{n}pa-sapa-wi* – "black-cherry moon."

No historical sources indicate any special harvesting techniques, so it is reasonable to assume that they were usually gathered by hand and placed in baskets for transportation. For taller trees, branches may have been pulled down or knocked with poles to reach the fruits. In the winter, pack rat nests were raided by the Comanche in order to obtain their plum caches.

To store the fruits, the seeds were removed, sometimes after the fruits were mashed, and the flesh was sun-dried, sometimes after being formed into cakes. This dried fruit was boiled to prepare it to eat, or the dried cakes were soaked in water. Boiling water was poured the dried fruit and the resulting infusion was drunk as a beverage (Haudenosaunee). The dried fruits were pounded in a mortar and mixed with dried, powdered meat to make a soup (Haudenosaunee).

The dried fruits were sometimes made into pemmican, or *wasna* (Blackfeet, Northern Cheyenne, Dakota, Sioux). A favorite dish of the Sioux, eaten at feasts, called "*wash-en-ena*" was made of pounded sun-dried *P. virginiana* fruits mixed with pulverized dried meat and marrow. This was eaten raw or cooked, but when cooked, meal made from the roots of *Pediomelum esculentum* was often added.

Seeds – Despite their toxicity, *Prunus* kernels were dried and stored for winter by the Comanche, and were boiled with fat to prepare them. The Cahuilla gathered *P. ilicifolia* seeds in August, sun-dried them, broke open the pits, and extracted the kernels, which were ground into meal in a mortar. The meal was then leached in a basket full of sand with cold water and boiled into a mush or beverage. The Cahuilla also ground up the seeds of *P. virginiana* and ate them as a meal, presumably after either leaching, cooking, or both. These cooking and leaching processes likely served to reduce or eliminate the cyanide content of *Prunus* kernels, but without testing or complete surety of methods used, this cannot be assumed to be safe.

Twigs – The twigs of *P. serotina* and *P. virginiana* were used for tea by the Ojibwe. They were made into a bundle about four inches long and one inch in diameter, tied together with a strip of bark. The strip was long enough to hold while dipping the bundle in hot water, and the resulting infusion was drunk as a beverage.

Bark – A decoction of *P. virginiana* bark was drunk as a beverage by the Menominee and Meskwaki.

Notes

WARNING – *Prunus* species, especially in their seeds and inner bark, contain dangerous concentrations of cyanogenic glycosides and over-consumption of the whole fruits or teas made from other plant parts can cause cyanide poisoning and death. Simply eating too many whole wild cherries, with seeds included, has caused death (Hidatsa).

Cyanide is present throughout the plant, but its concentration in the fleshy fruit mesocarp and exocarp is insignificant. The kernels, however, contain much higher levels. The hydrocyanic acid content of the kernels are 15 mg/kg in sweet almond (*P. amygdalus*), 851 mg/kg in *P. armeniaca* (apricot), 715 mg/kg in *P. persica* (peach), and 1062 mg/kg in bitter almond (*P. amygdalus*).[14] Severe toxicity could result from the consumption of about 30 of any of these kernels (except for sweet almond) by an adult, with fewer needed for children.[14] Boiling or cooking *Linum usitatissimum* seeds eliminates over 90% of their hydrocyanic acid content (initially measuring 375 mg/kg).[14] Cooking, drying, and leaching can all reduce cyanide in plant foods, but none of these methods can be assumed to completely remove it or guarantee safety.

Character – This genus has bark on its small trunks and branches that becomes very recognizable with experience. It is glossy, with a kind of shiny, "banded," mottled appearance, and the bark is often peeling. The plants also often have short "pseudo-thorns," or projections derived from stems that are not sharply pointed and are one to several inches long.

The most common edible wild species in Austin are Chickasaw plum (*P. angustifolia*), Mexican plum (*P. mexicana*), and black cherry (*P. serotina*). Chickasaw plum and Mexican plum are both shrubs or small trees, whereas black cherry can grow into a large tree. Chickasaw plum is more shrubby and is clonal, forming dense stands. Mexican plum grows as a small tree, usually with a single, larger trunk. The leaves of Chickasaw plum are more narrow and linear than the large, wide leaves of Mexican plum.

The Carolina laurelcherry (*P. caroliniana*) is a tree with glossy, thick leaves. Its fruits may not be edible due to their dryness or high cyanide content, though they are eaten by many birds and mammals. Peach (*P. persica*) is a small tree, most commonly cultivated but occasionally found wild in Austin. Its leaves are large, but more narrow and pointed than those of Mexican plum.

The ripe fruits of Chickasaw and Mexican plum are the largest, often larger than commercial sweet cherries (*P. avium*), and are very sweet, juicy, and soft, with a fruity flavor. They taste better than any commercial plum (*P. domestica*) available in stores. The ripe fruits of black cherry are smaller and somewhat bitter, but still sweet and rich in flavor. The flesh of the plums is yellowish whereas the flesh of the black cherry is dark red.

Prunus fruits are drupes, with a fleshy outer layer (mesocarp), thin skin (exocarp), and a single hard stone (endocarp). The stone acts as a protective shell for the inner kernel (seed), which resembles a commercial almond (*P. amygdalus*). In almond-type *Prunus* species, the mesocarp forms a green hull, similar to that of pecan, and the kernels are the part sought by dispersing animals.

Season – In Austin, Chickasaw plums ripen from late spring to early summer, peaking in the second half of May. Mexican plums ripen from midsummer to early fall, peaking in late August and early September. Black cherries ripen from late summer to mid-fall, peaking in the second half of August.

Nutrition – Fresh black cherries (*P. serotina*) contain 2% protein, 0% fat, 12% carbohydrate, and potassium (184 mg/100 g).[38] They are rich in phenolics and flavonoids, and have high antioxidant capacity, comparable to or exceeding that of commercial plums and grapes.[38]

Practice – Considering all the wild fleshy fruits in Austin, I rank the deliciousness of this genus as second or third, alongside dewberry; Texas persimmon comes first, and prickly pear fourth. I gather dozens of pounds of these fruits annually, especially in years with good Chickasaw or Mexican plum harvests. I gather far more of these than black cherries, which are less abundant, smaller, less delicious, and more difficult to gather. Nevertheless, black cherries remain an excellent foraging target when they are found in abundance.

I usually pick the fruits by hand, off the tree or from the ground. Fruits gathered from the ground must be inspected for integrity, and I avoid collecting those that have been sitting there for more than a week. In productive stands of Chickasaw plums during a good year, I can gather a dozen pounds of ripe fruits in about an hour.

I eat a lot of these fresh and raw, usually after a quick rinse. Preserving them using only Paleolithic technology is difficult, as they take a long time to dry whole and may spoil before fully drying. Cutting them in half and removing the seeds helps somewhat, but they are so juicy and soft that they fall apart the moment the skin is cut and the flesh clings to the seeds. The watery pulp must be spread out to dry, making the process messy and challenging.

The way I prefer to process them is to cook them low and slow until they disintegrate. I then strain out the seeds and skins using a coarsely woven basket or a wire mesh basket. At this point, the pulp can be boiled and canned, or simmered until it thickens, then spread on a baking tray or dehydrator sheet and dried into delicious fruit leather.

Prunus angustifolia Marshall
Chickasaw plum

Cherokee plum, sandhill plum
Loc.: E, SE, N, & sparse S TX; uncommon in Travis Co.
Form: shrub, tree, up to 25-30 ft. tall; perennial.
Flowers: Feb-May (white).
Food: Comanche, Southeast Natives.

Prunus mexicana S. Watson
Mexican plum

= Prunus americana var. lanata, P. pensylvanica var. mollis
Bigtree plum, inch plum
Loc.: C, E, & sparse N & S TX; common in Travis Co.
Form: shrub, tree, up to 10-35 ft tall; perennial.
Flowers: Feb-Apr (white, pink).
Food: no records.

Prunus serotina Ehrh.
Black cherry

= Prunus virens, P. parksii, Padus rufula
Chokecherry, wild cherry, wild black cherry, rum cherry, *capulín, cerezo*
Apache: *dzedeyui* – "sour choke cherry," Arapaho: *bí:nonó:'o:óé* – "berry bush," Cherokee: *ta·yá* – "cherry" / *ta'yaèlùehí* – "cherry lowland" / *ta'yá inagehei'í* – "cherry lives on mountains," Hocąk: *nąpakwijanįk* – "tree cherry drunk," Menominee: *iwî'skîpi mîna'xtik* / *iwî'skîpi me'nûn*," Mohawk: *erıgo'a*, Onondaga: *e'i'*, Ojibwe: *okwe' mîn* – "worm from a fly's egg," Forest Potawatomi: *okwe'mînûn* – "grub-worm berry"
Loc.: E, W, & C TX; common in Travis Co.
Form: shrub, tree, up to 25-110 ft. tall; perennial.
Flowers: Mar-June (white).
Notes – *Prunus serotina* var. *eximia* is the variety in the Edwards Plateau (Hill Country), and it only reaches a maximum height of 50 ft.
Food: Arapaho, Haudenosaunee, Hocąk, Menominee, Ojibwe, Forest Potawatomi, Northern Ute, Texians.

Other *Prunus* spp. in Travis Co.:
Prunus caroliniana Aiton
Carolina laurelcherry
Loc.: E & C TX; common in Travis Co.

Prunus minutiflora Engelm.
Texas almond
Loc.: C, SW, & sparse S TX; rare in Travis Co.

Prunus persica (L.) Batsch
Peach
= Amygdalus persica, Persica vulgaris
Cherokee: *kwaná*, Haudenosaunee: *gai'däe' odji'yă'*, Hopi: *sipa'la*
Loc.: all TX (sparse W); common in Travis Co.; cultivated; introduced.
Form: shrub, tree; perennial.

Prunus rivularis Scheele
Creek plum
Loc.: C, NE, & NW TX; uncommon in Travis Co.

Additional notes:

It is not related to *Prunus*, but one species worth mentioning is *Malvaviscus arboreus* (wax mallow or Turk's cap). I have found no evidence of this native plant's historical use as food by Indigenous peoples. However, the entire plant is edible and palatable. The fruits are especially good to eat, being large, soft, fleshy, and sweet berries. The flowers are faintly sweet and fruity, and can be dried to make a hibiscus-like tea. The young foliage is mild, herbaceous, and faintly mucilaginous. It is found in E, S & C TX and is very common in Travis Co., both in the wild and as an ornamental. The foliage and other parts of various other local native species in the Malvaceae (mallow family) are also edible and palatable, including *Abutilon fruticosum*, *Allowissadula holosericea*, *Callirhoe* spp., *Malvastrum coromandelianum*, and *Pavonia lasiopetala*.

Malvaviscus arboreus. L to R: flower, foliage, fruit.

Prunus angustifolia. Base photo: fruiting plant in late May. Overlay, clockwise from top right: range map (GBIF); leaf (narrow compared to *P. mexicana*); glossy, banded bark with pseudo-thorns; seed inside stone (anatomically equivalent to almond nuts); ripe fruit in late May.

Prunus mexicana. Base photo: flowering plant in mid-February. Overlay, clockwise from top left: fruiting plant in late September; ripe fruit; glossy, banded bark; range map (GBIF); flowers.

Prunus serotina. Base photo: fruiting plant in late August. Overlay, clockwise from top left: range map (GBIF); ripe fruits; glossy, banded bark.

SALICACEAE – Willow Family

Populus deltoides W. Barton ex Marshall
Eastern cottonwood

= Aigeiros deltoides, Monilistus monilifera, P. sargentii, P. texana, P. wislizeni
Common cottonwood, valley cottonwood, Carolina poplar, necklace poplar, *alamo*
Apache: *tis*, Northern Cheyenne: *xamáa-hoohtsėtse* (large tree) / *métse(oʔo)* (young tree), Choctaw: *ête hesha kaklahashe* – "tree leaf noisy," Dakota: *wága chaⁿ*, Diné: *tʼiis*, Gosiute: *soʼ-o-pi*, Hopi: *pashihürpbe*, Kiowa: ya-*hee-hwai* / *äʼhiʔñ* / *ä-heeñ* – "principal tree," Lakota: *čaŋjaxu* – "chewing wood" / *wagačaŋ*, Omaha-Ponca: *maa zhoⁿ* – "cotton tree," Osage: *baka hi*, Pawnee: *natakaaru*, Tewa: *te*
Loc.: all TX except far S; common in Travis Co.
Form: tree, up to 100-180 ft. tall; perennial.
Flowers: Feb-Apr (yellow).

History

Fruits – The ball-like green fruits were chewed as gum by Isleta children.
Foliage – The leaf buds were used as chewing gum by the Chiricahua and Mescalero Apache and Diné.
Inner bark – The sprouting young shoots were peeled in the spring, and the inner bark was scraped off and eaten for its sweet taste and nutritional value (Northern Cheyenne, Teton Dakota, Kiowa). It was said to combat scurvy (Kiowa), suggesting it may be high in vitamin C.
Sap – Aphid secretions, called "honeydew," were gathered from the leaves and eaten as a sugar by the Gosiute. The gum exuding from the tree trunks was used for chewing by the Diné.

Notes

Character – This is one of the largest trees in Austin. It is common alongside rivers and other bodies of water and is also a frequent ornamental. Sycamore grows in similar habitats, but its bark is smooth and white with peeling gray sections, whereas cottonwood bark is rough and deeply furrowed. The specific epithet "*deltoides*" refers to the deltoid or triangular shape of the leaves, unlike the palmately lobed leaves of sycamore. In the wind, the leaves flutter and produce a sound like rainfall, similar to its congener quaking aspen (*Populus tremuloides*), due to the flattened petioles characteristic of this genus.
Season – The leaf buds are available in late winter and early spring. The inner bark is best harvested in the same season. Honeydew occurs on only some trees, and is typically found in spring or summer.
Practice – The mature fruits are full of dry fluff that carries the seeds on the wind, giving the tree its name. It can be challenging to find many leaf buds within reach for harvesting. Both the buds and the inner bark are mildly sweet and bitter.

The honeydew from the leaves is sweet, but is difficult to extract. The leaves can be soaked in water to make a mildly sweet liquid. Historically, honeydew was extracted from *Phragmites* foliage by drying the leaves, beating them with sticks over a hide to dislodge the sugar crystals, and forming the crystals into balls.

Populus deltoides. Base photo: medium-sized tree. Overlay: range map (GBIF); bark; leaves (note their triangular or "deltoid" shape).

SAPOTACEAE – Sapodilla Family

Sideroxylon lanuginosum Michx.
Gum bumelia

= Bumelia lanuginosa, Lyciodes lanuginosum
Gum bully, coma, chittamwood, woolly buckthorn, black-haw, false buckthorn
Kiowa: *'ko-la* – "gum"
Loc.: all TX except NW; very common in Travis Co.; 3 *Sideroxylon* spp. in TX.
Form: shrub, tree, up to 50 ft. tall; perennial.
Flowers: June-July (white).

History

Fruits – The black fruits were sometimes eaten by the Kiowa, other Indigenous peoples in the Southern US, and folk in Northern Mexico.

The Seri ate the similar fruits of *Sideroxylon occidentale*. They crushed the fruits between their fingers to remove the seeds and ate the flesh raw. The fruits were dried, though not stored. They were sometimes gathered by raiding pack rat (*Neotoma*) nest caches.

Bark – The Kiowa ground up the outer bark (likely with the inner bark included) to produce a mucilaginous substance that was very soft, adhesive, and hardened rather quickly in the air. This was one of their favorite chewing gums.

Notes

Character – This tree can easily be mistaken for a live oak, but it has thorns. The small black fruits are another obvious characteristic.

Season – The fruits ripen from midsummer to early fall. The sap is available year-round, except from late fall to late winter. The young leaves are available in early spring.

Practice – I gather the fruits by hand or with the "strike and catch" method. They taste sweet and juicy, with a faint medicinal note. They vary in quality between trees, with some being more bitter or less sweet and juicy. I spit out the seeds when eating them fresh, but eat the fruits whole after drying, though I cannot confirm that the seeds are considered edible.

The fresh sap (white latex) has a clean, sweet taste. Once dried, it makes an excellent chewing gum with a faint sweetness and almost no other flavor. This kind of gum has the consistency of a quality commercial chewing gum but does not disintegrate or harden; it maintains the same texture no matter how long it is chewed.

The Kiowa account of using the outer bark may be erroneous. I have tried that technique a few times, yielding nothing but the expected result of grinding and chewing bark: very fibrous, distasteful, and seemingly pointless. There may have been more to the method, perhaps as a technique to extract the sap, or the authors may have misunderstood what they were told.

The best way to extract the gum is to make a slanting gash, with the lower end directed into a collection vessel. The easiest way to obtain the gum is to simply look for places where it has naturally extruded from the tree and dried in balls and drips.

The young, newly emerging spring leaves also make an excellent fresh snack in the field. They are somewhat chewy and fibrous, but have a pleasantly sweet and herbaceous taste. This observation is based on my personal experimentation, and I have not seen this use described elsewhere.

Additional notes:

It is not related to *Sideroxylon*, but another thorny shrub or tree worth mentioning is *Condalia hookeri* (Brasilian bluewood or Brasil). I have found no evidence of this native plant's historical use as food by Indigenous peoples. However, the similar fruits of *C. spathulata* were eaten fresh and raw by the Mexican Kickapoo and Tohono O'odham. The fruits of *C. hookeri* are small, red to black, and sweet and palatable. They ripen from fall through winter. The species is found in S, SW, & C TX and is common in Travis Co.

Condalia hookeri (Rhamnaceae) fruiting in mid-January. Note the prominent thorns, spoon-shaped leaves, and red to black ripe fruits.

Sideroxylon lanuginosum. Base photo: stand of shrub-sized plants in mid-April (note grayish cast to the foliage). Overlay, clockwise from top left: ripe fruits in late August; dried sap perfect for chewing gum in early August; young leaves in early April; range map (GBIF); ripe and unripe fruits in late August; latex from fresh wound in mid-September.

VINES

ROSACEAE – Rose Family

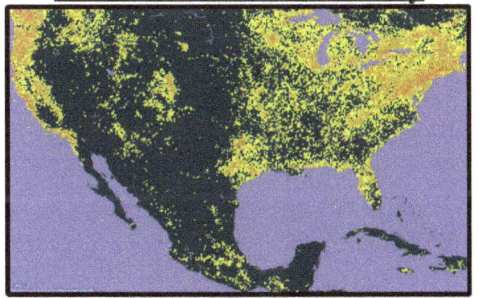 

Rubus distribution. Source: GBIF.

Rubus spp.
Blackberry

Cahuilla: *pikwlyam*, Comanche: *panatsayaa?*, Haudenosaunee: *otgä'ashä'*
Loc.: all TX except NW & far N; very common in Travis Co. (1 sp.); 16 spp. in TX (1 introduced).
Form: herb, low shrub, shrub, vine; perennial.

History

Fruits – *Rubus* fruits were eaten by the Chiricahua and Mescalero Apache, Asinai, Blackfeet, Caddo, Cahuilla, Calpella, Cherokee, Northern Cheyenne, Choctaw, Comanche, Concow, Cree, Dakota, Diné, Flathead, Gosiute, Haudenosaunee, Hocąk, Isleta, Karankawa, Karuk, Klamath, Little Lake, Northern Maidu, Menominee, Meskwaki, Nomlaki, Ojibwe, Omaha, Pawnee, Pomo, Ponca, Potawatomi, Shasta, Tolowa, Ute, Wailaki, Yokia, Yuki, Yurok, Natives in the Pacific Northwest, and other Natives. Species eaten include *R. allegheniensis*, *R. arcticus*, *R. argutus*, *R. canadensis*, *R. flagellaris*, *R. frondosus*, *R. glaucifolius*, *R. hispidus*, *R. idaeus*, *R. leucodermis*, *R. occidentalis*, *R. odoratus*, *R. nutkanus*, *R. pubescens*, *R. spectabilis*, *R. trivialis*, *R. ursinus*, and *R. vitifolius*.

Blackberries were commonly dried for winter storage. The preferred drying method of the Haudenosaunee was to break the stems bearing the fruits and allow them to hang on the plant to sun-dry. Presumably, a section of the stem bark was left connected, but all conductive tissues linking to the fruits were severed. This may have aided drying through wicking action. It also reduced the risk of microbial contamination that could result from handling the fruits directly. The fruits are fragile, so avoiding direct contact also helped prevent damage and loss of juices.

To make a beverage that was commonly drunk, the dried fruits were soaked in honey and water, or the fresh fruits were mixed with water and maple sugar (Haudenosaunee). Underripe *R. nutkanus* fruits were soaked in water for a beverage (Cahuilla)

Around 1530, Natives (likely Karankawa) of San Luis Island, next to Galveston, Texas, spent the month of April eating large quantities of *R. trivialis* fruits. It was a month of celebration and dancing, and large amounts of Black Drink (yaupon tea) were drunk. For the Choctaw, June was *hàsh bissa* – "blackberry month."

Foliage – The fresh shoots of some *Rubus* species were peeled and eaten (Haudenosaunee). The shoots of *R. idaeus* were peeled and the inner part was eaten (Cree) or made into a sweetened infusion for a beverage (Haudenosaunee). The young shoots of *R. nutkanus* and *R. spectabilis* were eaten fresh or steamed in bundles on hot stones (Tolowa, Northwest Natives). An infusion of the leaves and vines of *R. idaeus* was drunk as a beverage (Hocąk, Ojibwe). An infusion of the young leaves or root bark of *R. occidentalis* was drunk as a beverage (Dakota, Meskwaki, Omaha, Pawnee, Ponca)

Notes

Character – *Rubus* species are easy to identify by their fruits, which are aggregates, like mulberries, and resemble commercial blackberries. The plants are typically sprawling vines that form dense thickets low to the ground, although they can sometimes be found singly or climbing trees. The stems and petioles are covered with prickles that are straight to slightly curved. The leaves are palmately compound, with three, five, or seven leaflets. *R. trivialis* is the only wild *Rubus* species in Travis County, and is very widespread. They occur in many habitats but grow best in moist soils. The genus is complex and speciose, with about 230 species in North America north of Mexico, including about 12 introduced species.

Season – In Austin, ripe southern dewberries peak in availability in the second half of April and can be found from late March through mid-May.

Nutrition – *Rubus* fruits (raspberry and blackberry) contain about 1% protein, 1% fat, 11% carbohydrate, 6% fiber, potassium (157 mg/100 g) and vitamin C (24mg/100 g),[13] making them a good source of vitamin C. Fruits from various *Rubus* species are rich in anthocyanins, polyphenols, flavonoids, vitamins, and minerals, and they exhibit high antioxidant activity as well as anti-inflammatory, anti-neurodegenerative, and anti-cancer effects.[13]

Practice – Southern dewberries are one of my favorite foraging targets, and I would rank their deliciousness second only to Texas persimmons.

I scout for large patches in soft, moist soil. Once the fruits ripen, I gather them by hand. I usually wear high boots for comfort while carefully navigating the thorny thickets. Gardening gloves and long sleeves may also be recommended, though I typically work bare-handed and endure many prickles in my hands. A gentle touch is best: using three or four fingers, I grasp the entire fruit to pull it off, as they are prone to falling apart.

They are amazing fresh and raw, tasting very sweet, juicy, fruity, soft, and better than any blackberry that I have bought. I also cook them down to make excellent preserves and juice for canning. They can also be dried for storage.

The flowers are edible, and I have used them as a beautiful garnish for salads and desserts. I have also made various teas from the foliage, especially infusions of the young leaves. These teas are slightly bitter and astringent, but generally pleasant, and likely nutritious. I have also eaten the young shoots of southern dewberry both raw and cooked, though they are somewhat fibrous and bitter.

Rubus trivialis Michx.
Southern dewberry

= Rubus riograndis
Loc.: C, E, & S TX; very common in Travis Co.
Form: low shrub, vine; perennial.
Flowers: Mar-Apr (white, pink).
Food: Cherokee, Karankawa.

Additional notes:

The plants most commonly historically used by Indigenous North Americans to treat diarrhea and related ailments were plants in the genus *Rubus*, which includes dewberry, raspberry, thimbleberry, and salmonberry.

The roots of at least nine *Rubus* species were used by at least fifteen cultures for diarrhea, dysentery, and other gastrointestinal ailments (Cahuilla, Cherokee, Concow, Haudenosaunee, Little lake, Meskwaki, Nuxalk, Ojibwe, Omaha, Pomo, Shasta, Tolowa, Wintu, Yokia, Yuki). Species used include *R. allegheniensis*, *R. argutus*, *R. leucodermis*, *R. flagellaris*, *R. occidentalis*, *R. odoratus*, *R. spectabilis*, *R. ursinus*, and *R. vitifolius*.

I have tried this treatment on myself successfully. It is easy to loosen up the soil with a digging stick, pry out some roots, wash them, pound them, and make a decoction to drink. It is very bitter.

Rubus trivialis. Base photo: large fruiting patch in mid-April. Overlay, clockwise from top left: flower (pink or white); harvested fruits and young leaf; range map (GBIF); close-up of ripe fruit and stem prickles.

SMILACACEAE – Greenbrier Family

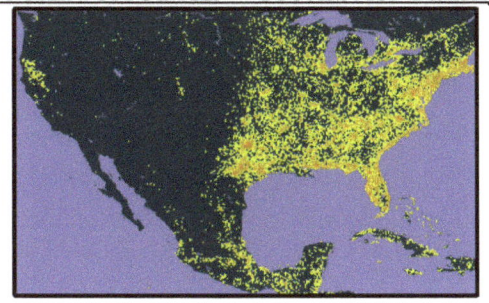 

Smilax distribution. Source: GBIF.

Smilax spp.
Greenbrier

Comanche: *tsuns*
Loc.: E, C, SE & NE TX; very common in Travis Co. (3 spp.); 10 *Smilax* spp. in TX.
Form: herb, vine, low shrub, shrub; perennial.

History

Fruits – The fruits of *S. herbacea* were eaten by the Meskwaki and Omaha, mainly for their taste.

Shoots – During the 1540 De Soto expedition, while starving in South Carolina, members reported having "nothing to eat other than the tendrils of young vines found growing in the woods and streams," most likely referring to the young shoots of *Smilax* species, which are quite palatable and common in this area "full of brambles." As they were accompanied by Indigenous guides at the time who foraged other "edible herbs and roots", they likely learned this food source from them.

Roots – Greenbrier roots were eaten by the Choctaw, Comanche, Houma, and Southeast Natives. Species eaten include *S. auriculata*, *S. bona-nox*, *S. laurifolia*, *S. pseudochina*, and *S. rotundifolia*.

The tuberous rootstocks of *Smilax* were used by Southeast Natives for food, with their starches being made into a soup, bread, or jelly. The starch was "readily obtained as a reddish sediment by washing in water." They likely pounded up the roots, soaked them in water, and used the starches that settled on the bottom of the vessel after removing the fibrous parts and pouring off the water.

The Comanche dug greenbrier roots from deep underground, removed their thin bark, and roasted the inner part. The Choctaw peeled the roots, sliced and dried the inner part, then ground the dried slices into a powder that was baked into a kind of bread. They also pounded up the roots of *S. laurifolia* (called "*ahe*" or "*kanták*") into a fine meal, mixed it with water to make a paste, and formed this into small cakes, which, when fried in grease, were said to be one of their favorite foods. The Houma dried the roots of *S. bona-nox* before pounding them into meal in a mortar and pestle, using it in a manner similar to cornmeal. They likely processed them in a more detailed manner to remove the fibrous parts, especially the outer layer.

Greenbrier (possibly *S. bona-nox*) was historically called *zarzaparilla*, in reference to its root being used to prepare the flavorful beverage sarsaparilla, for which several other species in this genus are known.

Notes

Character – These vines are very common and easily recognizable by their green, prickly (thorny) stems. Depending on the species, the growing tips, including the tendrils, leaves, stem, and prickles are soft enough to eat for a length of about four inches to a foot. Past this point the plant becomes fibrous and, while still technically edible, cannot be chewed, especially the hard prickles. The fruits are borne in spherical clusters, each containing anywhere from a few to about two dozen dark purple to black fruits. Each fruit contains a single large seed surrounded by a gelatinous layer, a fleshy layer, and a thin skin.

S. bona-nox is the most common species in Austin. It has widely spaced thorns and smaller leaves (1.5-4.5 in. long). Its leaves are usually lobed to some degree and often exhibit variegation or whitish mottling. It has thorns on its stems opposite its petioles, and often has prickles along its leaf margins. *S. rotundifolia* is very similar, but its leaves (2-6 in. long) are often more rounded, it lacks thorns opposite the petioles, and it never has prickles along the leaf margins. *S. tamnoides* has the largest leaves, sometimes reaching nearly a foot in diameter, and the thickest stems. It can be distinguished by the very dense distribution of thorns along its stems, giving it the name "bristly greenbrier."

Season – The fruits are available from fall through winter. The growing tips are most abundant in the spring, but can be found year-round. There is likely an ideal gathering season for the roots, perhaps fall or early spring, but they are also available year-round.

Practice – The tender, young growing tips are mild-flavored, crisp, and pleasant to eat. I have eaten these raw thousands of times over decades, as this was the second wild edible green I learned as a child (after woodsorrel). Oddly, there are no explicit mentions of this food use in the ethnobotanical sources I have consulted. Perhaps it was seen as a mere opportunistic snack rather than a viable food source, known but neglected in the records.

The roots are difficult to excavate, especially in rocky soils of Austin. They are very tough and fibrous and must be pounded up and soaked in water to release the starches, which settle to the bottom and can be collected after removing the fibrous parts and pouring off the water. Despite the labor involved, under the right conditions they can furnish a significant amount of food. Roots of *S. tamnoides* growing in softer soils are probably the best target for foraging, though the roots of all *Smilax* species can be used. Some roots have bulbous portions, which may contain a higher concentration of starch.

The fruits are mildly sweet and faintly astringent. The gelatinous layer is difficult to remove from the seed, but applying the right bite pressure can separate it. They make an interesting snack, though they are not a significant food source.

Smilax bona-nox L.
Saw greenbrier

Bullbrier, catbrier, stretchberry vine, *zarzaparilla*
Choctaw: *kanták*, Houma: *kanták / cantaque*
Loc.: all TX except far W & N; very common in Travis Co.
Form: herb, vine; perennial.
Flowers: Mar-May (green).
Food: Choctaw (roots).

Smilax rotundifolia L.
Roundleaf greenbrier

Bullbrier
Cherokee: *aniskiná unanesadá* – "for the leg," Kiowa: *moñ-ksoñ-a-a* – "sharp grass"
Loc.: E & C TX; not uncommon in Travis Co.
Form: herb, vine; perennial.
Food: Southeast Natives.

Smilax tamnoides L.
Bristly greenbrier

Bristly sarsaparilla
Choctaw: *beshu'kchenokle*, Hocąk: *waxacsep* – "black stickers"
Loc.: E, SE, NE, & C TX; common in Travis Co.
Form: vine, shrub; perennial.
Food: no records.

Additional notes:

Smilax bona-nox leaves were used by the Comanche to roll tobacco for smoking.

Smilax bona-nox. Base photo: small plant. Overlay, clockwise from top left: ripe fruits in early December; leaf close-up (note prickles along leaf margin); range map (GBIF); harvested and cleaned roots.

Smilax rotundifolia. Base photo: foliage (note lack of prickles along the leaf margin and lack of thorn opposite the petioles). Overlay, clockwise from top left: range map (GBIF); ripe fruits in late December; harvested young tips of *S. rotundifolia* and *S. bona-nox* in late March.

Smilax tamnoides. Base photo: foliage (note large leaves). Overlay, clockwise from top right: range map (GBIF); young growing tip in early August; densely thorny stem; ripe fruits in late November.

VITACEAE – Grape Family

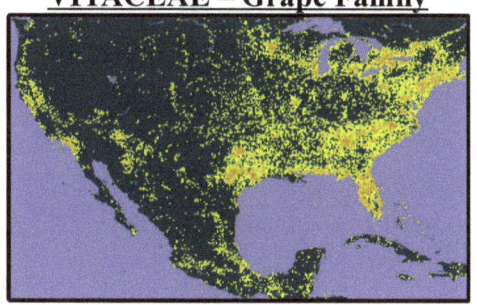

Vitis distribution. Source: GBIF.

Vitis spp.
Grape

Comanche: *natsomukwe / nosinatsomukwe*, Haudenosaunee: *oñiŭng'wisä'*
Loc.: all TX; very common in Travis Co. (5 spp.); 13 spp. in TX.
Form: shrub, vine, up to 50 ft. long; perennial.

History

Fruits – Wild grapes were eaten by the Acansa, Apache, Asinai, Caddo, Cahuilla, Calpella, Cenis, Cherokee, Northern Cheyenne, Comanche, Concow, Dakota, Diné, Haudenosaunee, Hocąk, Isleta, Jemez, Karankawa, Kiowa, Little Lake, Menominee, Meskwaki, Natchitoches, Nomlaki, Ojibwe, Omaha, Pawnee, Pomo, Ponca, Pueblo, Quapaw, Wailaki, Winnebago, Yokia, Yuki, Yurok, Natives in Florida, South Carolina, Arkansas, and Kansas in 1540, Natives in South Carolina in 1567, Natives from California to Arkansas, Texas Natives 5,000 years ago, Natives in south-central Texas in 1709, and likely others. Species eaten include *V. aestivalis*, *V. arizonica*, *V. californica*, *V. cinerea*, *V. girdiana*, *V. labrusca*, *V. mustangensis*, *V. riparia*, and *V. vulpina*.

Great quantities of wild grapes were gathered and stored (Caddo, other Natives). They were eaten fresh or were sun-dried for storage (Apache, Caddo, Cahuilla, Comanche, Dakota, Diné, Isleta, Jemez, Karankawa, Kiowa, Meskwaki, Omaha, Pawnee, Ponca, Winnebago).

Among the Comanche, dried grapes were mashed, mixed with grease, and stored for winter, with chunks toasted before eating. The dried fruits were also moistened, worked into cakes, and baked, with fat sometimes added. Dried grapes were boiled to prepare them for eating (Cahuilla, Comanche). The resulting decoction was also used by the Comanche to make grape dumplings.

In the 1680s, the Cenis were said to have white grapes, which may have been a special variant of a native Texas grape species that they cultivated or tended. Among Puebloan peoples, *V. arizonica* vines were cultivated in rows and were an important food source from the time of the the first Spanish expeditions, they were likely cultivated in the pre-contact period.

Shoots – The fresh, unpeeled shoots of *V. vulpina* were eaten by the Haudenosaunee.

Sap – Large grapevines (*V. cinerea* and *V. vulpina*) were tapped in the spring to collect sap, which was drunk fresh and was said to taste like the juice of the fruits (Dakota, Kiowa, Omaha, Pawnee, Ponca, and Winnebago).

Notes

Character – Grapevines are one of the most common elements of forests in Austin, especially in riparian habitats. Mustang grape is by far the most prevalent species of grape in the area.

Vitis mustangensis can be easily distinguished from the other species by the whitish, densely pubescent underside of its leaves. Its leaves may be entire or highly divided into lobes. *V. berlandieri* and *V. cinerea* are not uncommon in the area and can be distinguished by leaf size and shape. *V. cinerea* leaves are 4-8 inches long or broad, with rough, uneven serrations along the margins. *V. berlandieri* leaves are 2-6 inches broad, with even serrations along the margins and an acutely tapering leaf tip. Both species may have lobed leaves. *V. berlandieri* prefers limestone soils, whereas *V. cinerea* prefers alluvial soils. *V. monticola* is also not uncommon and has markedly glossier leaves than the above species, with evenly serrate margins. *V. aestivalis*, *V. rupestris* and *V. vulpina* also occur in Travis County, but are rare. *V. aestivalis* has deep, widely spaced lobes. *V. rotundifolia* (muscadine) does not occur in Travis County.

Ampelopsis cordata is in the grape family and closely resembles grape vines, but is not edible. Its fruits are distinct from grapes, having a blue to pink color with small spots. The foliage lacks the tart, palatable taste of grape leaves and instead is bitter and unpleasant. Its bark is bumpy rather than roughly fissured. *Nekemias arborea*, also in the grape family, produces fruits that resemble grapes but are glossier and inedible. Its leaves, unlike grapes, are compound, with serrate margins.

The bark of the large stems of grapevines has deep longitudinal fissures. These vines are remarkably flexible, even when several inches in diameter. Grapevines were historically used by Native peoples for lashings. A grapevine (likely mustang grape) with a diameter of 8 feet was found by a 1716 Spanish expedition at San Miguel Creek in Frio County.

As a fascinating aside, the evolutionary center of origin for *Vitis* is

North America.[78] The closest relative of *V. vinifera*, the commercial grape species, is *V. californica*.[78] It is no coincidence that California grows the vast majority of grapes in the US and ranks among the top grape-producing regions globally. *V.* x *champinii*, native to the Edwards Plateau (Hill Country), is used as a rootstock in modern commercial grape production and was likely produced by hybridization between *V. mustangensis* and *V. rupestris*.[78] *V. berlandieri*, another grape native to the Edwards Plateau, helped save the French wine industry from the grape phylloxera because of its resistance to these aphids and is used as a rootstock in French wine cultivars.[28] The Edwards Plateau was uniquely important in the evolution of grape species.[78] The first Old World explorers known to visit the New World, around 1000 AD, referred to North America as "Vinland" because of the abundance of wild grapes, according to the Vinland Sagas.

Season – Ripe wild grapes are most abundant in Austin during June and July, peaking in mid-July for mustang grapes, though they can occasionally be found as late as September.

Nutrition – Fresh, seedless, cultivated red grapes (*V. vinifera*) contain 1% protein, 0% fat, 20% carbohydrate, and potassium (229 mg/100 g).[69] Wild grapes have higher protein and fat content because they contain seeds. They also likely have higher concentrations of vitamins and minerals and lower sugar content, as cultivated grapes have been bred for sugar quantity rather than nutritional quality.

Practice – The foliage of *Vitis* species is edible and tastes tart, herbaceous, and mildly sweet. I often eat the leaves raw, especially when young. Mature mustang grape leaves are rather fibrous, but I like to chew them up, extract the juices, and discard the fibrous material. The flowers and young stems also make a tasty snack.

Of all the plants producing fleshy wild fruits in Austin, I can gather the most from mustang grape vines. The challenge of grape foraging is finding fruiting vines close enough to the ground. The vines can often be found cascading from the forest canopy to the ground at edges of rivers and creeks, as well as on fences or trees bordering open areas. I have identified a number of such prime locations, from where I harvest every summer. I gather them by hand, pulling off entire clusters of grapes rather than trying to pick them one by one. When I grasp a single grape and pull, usually only the skin and outer flesh come off, leaving the central part on the stem. Instead, I grasp the base of the whole cluster and pull it perpendicularly off the stem, then put it in my bag. During peak season, I can easily gather a dozen gallons of grapes in a few hours. This has practically no impact, as they are produced extremely abundantly and most remain out of reach.

To process them, I generally dry or juice the grapes. To dry them, I remove each grape from its stem and place them on a drying tray in a single layer to dry in the sun, oven, or dehydrator. Removing them from the cluster speeds drying, as it leaves a hole for moisture to escape. It is also difficult to cleanly detach the grapes from the stems once they are dried. Sun-drying takes about a week or two, a dehydrator a few days, and an oven about half a day. The dried grapes are not like store-bought raisins, which typically have added sugar; they are somewhat crisp and sour. I think they make an excellent snack, which I eat whole and plain year-round or add to smoothies. They can also be rehydrated for recipes or boiled to make a decoction that is similar to juice.

To juice the grapes, the easiest method that I have found is to put entire clusters into a blender and blend them until smooth, sometimes adding water. Then I strain the mixture through a wire mesh strainer or basket. The strainer will need to be shaken a bit or intermittently emptied and cleaned as the pulp will clog the pores and slow drainage. I like to use two strainers: as the flow through the first slows, I dump the pulp into the second, rinse the first, and repeat the process. Using a nylon or cotton bag is possible but much slower and requires considerable force to squeeze and knead out the juice. Once the pulp forms a solid mass, I compost it and heat the juice to a low simmer. The juice can then be boiled and canned or drunk. I usually add about half water and sometimes a bit of sugar to make juice, which is otherwise rather concentrated.

Another method I use to can or juice them is to remove them all from the stems and cook them in a pot on a low simmer for a while. Once softened and disintegrating, I use a potato masher to completely pulp them in the pot. The pulp can be boiled and canned straight, but the product is superior if I strain out the seeds and skins, which I do as described above.

Some wild grapes, including mustang grape, can cause a burning sensation in the mouth or on the skin. This does not occur from all individuals but is a common trait. Fully ripe grapes are less likely to produce this effect. I sample from different vines and ensure they are fully ripe to prevent this. However, grapes that cause this reaction are perfectly edible, even raw, despite the unpleasant sensation. The burning property can also be easily eliminated by drying or cooking the grapes, which destroys the compound responsible.

Vitis aestivalis Michx.
Summer grape

Arkansas grape
Cherokee: *te·lo dí* / *telaładí* / *telà·ladí* – "it hangs down"
Loc.: E & sparse C & S TX; rare in Travis Co.
Form: vine; perennial.
Flowers: Apr-June (white, yellow, green).
Food: Cherokee.

Vitis berlandieri Planch.
Berlandier's grape

= V. cinerea var. helleri
Loc.: C & sparse E, N, & S TX; not uncommon in Travis Co.
Form: vine; perennial.
Food: no records.

Vitis cinerea (Engelm.) Millardet
Graybark grape

= Vitis berlandieri (= V. cinerea var. helleri), V. cordifolia, V. helleri
Ashy grape
Dakota: *hasta^{n}ha^{n}ka*, Teton Dakota: *cha^n wíyape* – "tree twine," Meskwaki: *sîwanûnj^h* / *siwa'nûn*, Omaha-Ponca: *hazi*, Pawnee: *kisúts*, Winnebago: *hapsintsh*
Loc.: C, E, SW, & sparse S & N TX; not uncommon in Travis Co.
Form: vine; perennial.
Flowers: May-July (yellow, green).
Food: Dakota, Kiowa, Meskwaki, Omaha, Pawnee, Ponca, Winnebago.

Vitis monticola Buckley
Sweet mountain grape

Champin grape
Loc.: C & SW TX; not uncommon in Travis Co.
Flowers: May-June (white, green).
Form: vine, stems up to 40 ft. long; perennial.
Food: no records.

Vitis mustangensis Buckley
Mustang grape

= Vitis candicans
Loc.: C, E, & S TX; very common in Travis Co.
Form: vine, stems up to 50 ft. long; perennial.
Flowers: Apr-June (white, red).
Food: Natives in south-central Texas in 1709.

Vitis mustangensis. Base photo: fruiting vine in late June. Overlay, clockwise from top left: underside of leaf (note dense, whitish pubescence); harvested grape clusters in late June; flowers in late March; bark of large vine stems; dried fruits; range map (GBIF).

Top: *Vitis berlandieri*. Note the acutely tapering leaf tip and more even serrations.
Bottom: *V. cinerea*. Note the larger size, rougher serrations, and lack of tapering leaf tip.
Overlay, left: *V. monticola*. Note the smooth glossiness.
Overlay, right: *V. aestivalis*. Note the wide spacing in the deepest part of the lobes.
Range maps (GBIF), clockwise from top right: *V. berlandieri, V. aestivalis, V. cinerea, V. monticola*.

NOT EDIBLE LOOKALIKES
Top: *Ampelopsis cordata* (heartleaf peppervine) foliage, fruits, and range map (GBIF). Note the fruit color and bark texture.
Bottom: *Nekemias arborea* (peppervine) foliage, fruits, and range map (GBIF). Note the compound leaves.

HERBS

DICOT HERBS

AMARANTHACEAE – Amaranth Family

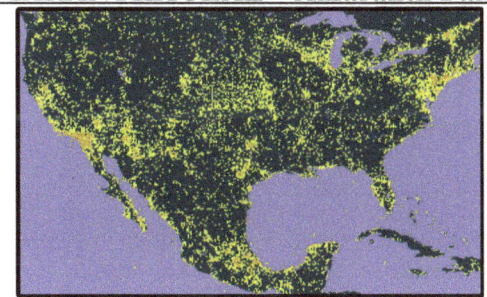

Amaranthus distribution. Source: GBIF.
***Amaranthus* spp.**
Amaranth

Pigweed, *quelite*
Gosiute: *ats*
Loc.: all TX; common in Travis Co. (10+ spp.); 25 spp. in TX (5 introduced).
Form: herb, low shrub; annual.

History

Seeds – The seeds of various *Amaranthus* species were eaten by the Apache, Caddo, Calpella, Cahuilla, Concow, Diné, Gosiute, Hopi, Kiowa, Klamath, Little Lake, Nomlaki, Pomo, Seri, Wailaki, Yokia, Yuki, Zuñi, Montana Natives, and Western Natives. Species eaten include *Amaranthus albus*, *A. blitoides*, *A. cruentus*, *A. fimbriatus*, *A. hypochondriacus*, *A. palmeri*, *A. retroflexus*, and *A. watsonii*.

The seed heads (infructescenses) were cut or broken from the plant and gathered in a container (Apache). The seeds were isolated by rubbing the seed heads between the hands (threshing) over buckskin, cloth, or a basket, then winnowing off the debris (Apache, Cahuilla, Seri). Whole plants were also gathered and dried before threshing (Western Natives). Seeds were sometimes parched (Diné, Tohono O'odham, Seri) or eaten raw (Zuñi). They were parched by shaking them in a basket with live coals (Tohono O'odham).

The seeds were ground into a meal (Apache, Cahuilla, Diné, Hopi, Kiowa, Seri). This meal was eaten as piñole (Diné, Kiowa), as a drink (Cahuilla), as a mush (Cahuilla, Diné, Seri), or formed into cakes or bread (Apache, Cahuilla, Zuñi).

Amaranth seeds were produced in abundance and were an important food source (Apache, Gosiute, Seri, Tohono O'odham, Western Natives). The seeds were sun-dried and stored (Diné, Hopi, Kiowa, Tohono O'odham, Seri), sometimes kept in sealed pottery jars (Seri). *Amaranthus* species (including *A. cruentus*, *A. hypochondriacus*, and *A. retroflexus*) were cultivated (Southern Paiute) or allowed to volunteer in gardens (Hopi, Round Valley Natives) for use as food.

The seeds were gathered in September (Tohono O'odham).

Foliage – The foliage of various amaranth species was eaten by the Apache, Cahuilla, Diné, Hopi, Isleta, Jemez, Kiowa, Southern Paiute, Seri, Tewa, Tohono O'odham, Montana Natives, and Western Natives. Species eaten include *Amaranthus acanthochiton*, *A. albus*, *A. blitoides*, *A. cruentus*, *A. fimbriatus*, *A. graecizans*, *A. hypochondriacus*, *A. palmeri*, *A. retroflexus*, *A. torreyi*, and *A. watsonii*.

The aboveground parts were eaten, especially the young leaves and shoots. The plants were eaten boiled, fried, in soups, or in stews (Apache, Cahuilla, Diné, Hopi, Isleta, Jemez, Kiowa, Seri, Tewa, Tohono O'odham, Montana Natives). They were sometimes boiled, then fried (Diné, Kiowa, Seri, Tewa). The foliage was sometimes eaten raw (Apache, Isleta, Jemez).

The greens were an important food source of the Hopi, Kiowa, and Tohono O'odham. The foliage was hung in the house for storage (Hopi).

The foliage was gathered in the late summer (Cahuilla) or July and August (Tohono O'odham).

Notes

Character – Amaranth grains have been cultivated for more than 8,000 years in Mesoamerica and the Andes Mountains and are still commercially grown for grain and leafy vegetables, especially *A. caudatus*, *A. cruentus*, and *A. hypochondriacus*.[64]

Amaranthus species are herbaceous annuals, found throughout the US, with more species occuring in southern states. Some species form tumbleweeds. They can grow in low mats or reach up to six feet tall. The plants have simple, alternate leaves, usually ovate to lanceolate in shape, and are generally green, though some are reddish or grayish. The flowers and seeds are very small and densely packed on terminal spikes. They grow in many habitats and often thrive in disturbed areas. They are quite tolerant of drought and poor soils, with many being capable of growing even in urban asphalt habitats.

Nutrition – Amaranth grains contain 15-18% protein, 6-8% fat, and 60-

65% carbohydrate, and are rich in lysine, tryptophan, and sulfur amino acids.[51] The lysine and tryptophan content, as well as its poor leucine content, make it an excellent complement to maize, which is poor in lysine and tryptophan and rich in leucine.[51] Amaranth grain is rich in linoleic and palmitic acids.[51]

Amaranth aboveground parts (such as the foliage) contain 16-30% protein, 11-24% fiber, and 2-3% fat, depending on the growth stage.[51] Younger plants (80 days) have higher protein and lower fiber than older plants (120 days).[51]

Season – The foliage is available for most of the year, except during the coldest season. The flowering and fruiting seasons vary between species, and both can occur year-round, except usually from mid- to late winter. The seeds most commonly ripen from late summer through early fall.

Practice – The foliage of various amaranth species, especially when young, is quite palatable and serves as an excellent substitute for spinach and other leafy greens. It can also be dried for storage.

The seeds are best gathered by cutting off the seed heads just after the seeds fully develop, but before they have dried and dropped off. These seed heads can be dried, and then threshed over a fabric sheet by striking them with a stick or rubbing them by hand. The tiny seeds can then be gathered together and the debris winnowed off in a gentle breeze. These can be cooked and eaten like quinoa.

Amaranthus albus L.
Prostrate pigweed

= Amaranthus gracecizans, A. pubescens
Common tumbleweed
Loc.: all TX; uncommon in Travis Co..
Form: herb; annual.
Food: Apache, Western Natives.

Amaranthus blitoides S. Watson
Mat amaranth

Prostate amaranth, spreading pigweed, *quelite*
Diné: *naaskhaatiih* – "spread out," Hopi: *pociüh / posiüh / po:'siowu*, Klamath: *bä-lō'-ōch*, Tewa: *su*, Zuñi: *ku'shutsi* – "many seeds"
Loc.: all TX except far E; uncommon in Travis Co.; introduced.
Form: herb; annual.
Notes – the Hopi name refers to the black seeds, with "*posi*" meaning "seed."
Food: Diné, Hopi, Kiowa, Klamath, Zuñi, Montana Natives.

Amaranthus cruentus L.
Red amaranth

= Amaranthus hybridus var. paniculatus, Amaranthus paniculatus

African-spinach, coxcomb
Hopi: *komo / ko'mo*, Zuñi: *i'shilowa yäl'tokĭa* – "red, face paint"
Loc.: NE, C, & S TX; uncommon in Travis Co.; introduced.
Form: herb; annual.
Food: Hopi (especially an infusion of the whole seed head to dye corn bread a pink color).

Amaranthus palmeri S.Watson
Carelessweed

Palmer amaranth
Hopi: *komótoshu*, Tohono O'odham: *tcuhukia*
Loc.: all TX; common in Travis Co.
Form: herb; annual.
Food: Diné, Tohono O'odham.

Amaranthus retroflexus L.
Redroot pigweed

Redroot amaranth green amaranth, rough pigweed, *quelite, alegría*
Apache: *ndaji* – "black eye," Cayuga: *diunhe˙'gǫ*, Cherokee: *wats'ká*, Diné: *tł'óhtéesk'ítíh* – "grass, seeds, humped" / *tł'óhtéexócíh* – "grass, seeds, prickly," Jemez: *tſie fuoh iah*, Kawaiisu: *puguzivi̜*, Onondaga: *ganadanǫ˙'wi'*, Tewa: *su*
Loc.: all TX; uncommon in Travis Co.; introduced.
Form: herb; annual.
Food: Apache, Calpella, Concow, Diné, Little Lake, Nomlaki, Pomo, Wailaki, Yokia, Yuki, Western Natives.

Other *Amaranthus* spp. in Travis Co.:
Amaranthus blitum L.
Purple amaranth
Loc.: E, S, & sparse C TX; uncommon in Travis Co.; introduced.
Form: herb; annual.

Amaranthus polygonoides L.
Tropical amaranth
= Amaranthus berlandieri, A. taishanensis
Loc.: C, W, S, SE, & sparse E TX; common in Travis Co.
Form: herb; annual.

Amaranthus spinosus L.
Spiny amaranth, spring amaranth
Cherokee: *toletiyusti* – "stick on you, like"
Loc. E & S TX, sparse N TX; rare in Travis Co.

Amaranthus tuberculatus (Moq.) Sauer
Roughfruit amaranth
Loc.: C, N, E, & far S TX; uncommon in Travis Co.
Form: herb; annual.

Amaranthus viridis L.
Slender amaranth, pakai
Loc.: E, S, & sparse C TX; uncommon in Travis Co.; introduced.
Form: herb; annual.

Amaranthus palmeri. Top: flowering plants in late April. Bottom: fruiting plants in late July. Overlay: seed heads; range map (GBIF).

Top: *Amaranthus cruentus.* Left: fruiting plant in mid-August. Right: flowering plant in late November. Overlay: range map (GBIF).
Bottom: *Amaranthus blitoides.* Base photo: flowering plant in late November. Overlay: close-up of flowers; range map (GBIF).

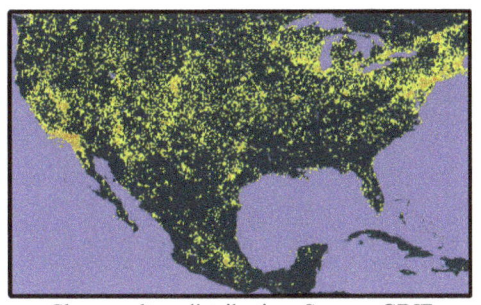 

Chenopodium distribution. Source: GBIF.

Chenopodium spp.
Goosefoot

Cahuilla: *ki'awet*, Diné: *tł'ohteeiłpáhíh* – "grass, seed, grey" / *tł'ohteeintł'ízíh* – "grass, seed, hard" / *tł'ohteeixwocíh* – "grass, seed, prickly," Haudenosaunee: *gwis'gwis ganĕ"das*, Hopi: *si'swa* / *hɜhɜ'la*

Loc.: all TX; common in Travis Co. (4 spp.); 23 species in TX (2 introduced).
Form: herb, low shrub, shrub, tree; annual, perennial.

History

Seeds – The seeds of various *Chenopodium* species were eaten by the Caddo, Cahuilla, Diné, Gosiute, Kiowa, Klamath, Seri, Zuñi, and other Natives. Species eaten include *Chenopodium album*, *C. californicum*, *C. fremontii*, *C. leptophyllum*, and *C. murale*.

The seeds were gathered by striking the plants with a basketry racket, casting the seeds downward into a waiting receptacle (Cahuilla). They were parched or dried, ground into a meal, and either baked into cakes or eaten as mush (Cahuilla, Diné, Kiowa, Klamath, Seri, other Natives). The Diné used them much like corn, making tortillas and bread, or simply parching and chewing them by the handful for their sweet taste. The Zuñi ground the seeds into meal, mixed it with cornmeal, seasoned the dough with salt, and formed it into balls or cakes, which were then steamed.

The seeds were gathered in large quantities and were an important food source, with granaries or sealed pottery jars being filled for storage (Cahuilla, Gosiute, Seri).

The seeds of *C. murale* were gathered in the late spring (Seri).

Foliage – The foliage of various goosefoot species was eaten by the Cahuilla, Calpella, Concow, Teton Dakota, Diné, Flathead, Haudenosaunee, Hocąk, Hopi, Isleta, Kiowa, Little Lake, Menominee, Nomlaki, Omaha, Pomo, Forest Potawatomi, Pueblo, Tohono O'odham, Ute, Wailaki, Yokia, Yuki, and Zuñi. Species eaten include *Chenopodium album*, *C. californicum*, *C. fremontii*, *C. humile*, *C. incanum*, *C. leptophyllum*, and *C. murale*.

The young plant parts (leaves and shoots) were eaten. They were eaten raw, cooked, boiled, fried, or included in stews. After boiling, the water was sometimes discarded to improve the taste.

The foliage was gathered in the summer or July and August (Tohono O'odham).

Notes

Character – Quinoa is a common commercial crop from the seeds of *Chenopodium quinoa*, which has been cultivated in the Andean region for 5,000 years.[54] *Chenopodium* species were also cultivated in Bavaria, Europe 3,000 years ago.[30]

Season – The foliage is best in spring and early summer, but can be eaten anytime it is still alive and green, which may be year-round. The seeds ripen from late summer through winter.

Nutrition – *Chenopodium* spp. seeds contain 11-14% protein and 3-6% fat.[53] The leaves of *Chenopodium* spp. contain 3-6% protein by fresh weight.[53] The younger leaves of *C. quinoa* are richer in protein and lower in oxalate compared to older leaves.[53] Vitamin C and carotenoids are most concentrated in middle-aged (middle of plant) leaves.[53]

The fresh (71% moisture) young leaves of *C. album* contain 9% protein, 6% carbohydrate, 1% fat, and 6% fiber.[30] *C. murale* and *C. opuulifolium* have similar nutrient compositions.[30] All three species are rich in vitamin C, carotenoids, and essential fatty acids.[30] The highest antinutritional component in these plants is oxalic acid, which can be reduced by cooking in water.[30] *C. album* is rich in bioactive phenolics with high antioxidant activities.[30]

Practice – The foliage of *C. album* and *C. giganteum* is among my favorite wild greens. It is highly palatable and nutritious. I eat it plain and raw, in salads, cooked like spinach, or blended in smoothies. It can also be dried for storage.

The seeds of *Chenopodium* are best gathered with the "strike and catch" method or by collecting whole seed heads just before they release the seeds, drying them, threshing out the seeds over a sheet, and winnowing off the debris. The seeds are very similar to quinoa and can be cooked in the same manner.

Chenopodium album L.
Lamb's-quarter

= Chenopodium alba, C. opulifolium, C. viride
Apache: *ita* – "leaf," Cayuga: *gwı́ ́sgwıs gadiwano ́gras* – "pig eats it," Cree: *wīthiniwpakwātik*, Dakota: *waȟpe toto* – "greens," Diné: *tł'ohteei'tshoh* – "grass, seeded, big," Hocąk: *raxgemąkejahağep* – "old ground growing," Hopi: *cirswa* / *sü'rswa* / *si'swa* / *hɜhɜ'la*, Kiowa: *bàtl-sai-añ* – "stink-weed" / *bàdl-sai-ya-doñ*, Lakota: *čanxloǧan iŋkpa gmigmela* – "small end (of the leaves?) rounded weed," Meskwaki: *tcakû'skį^h* – "stickers" / *askipwawis äpīsane kwäyāwigi*, Mohawk: *skanadanųwı ́*, Onondaga: *ganadanǫ ́ ́wi ́*, Pawnee: *kitsarius* – "green juice," Forest Potawatomi: *koko'ʃîbag* – "pig leaf"
Loc.: all TX; common in Travis Co.
Form: herb; annual.
Flowers: June-Oct. (green, brown).
Notes – There are conflicting data about its introduced status according to WFO, BONAP, USDA, POWO, and other plant databases. It is native to North America, but in Texas, it may be native (BONAP, USDA) or introduced (POWO, WFO).
Food: Calpella, Concow, Teton Dakota, Diné, Haudenosaunee, Hocąk, Hopi, Kiowa, Little Lake, Menominee, Nomlaki, Omaha, Pomo, Forest Potawatomi, Pueblo, Tohono O'odham, Ute, Wailaki, Yokia, Yuki, other Natives.

Chenopodium giganteum D. Don
Tree spinach

= Chenopodium album ssp. amaranthicolor, C. album var. album, C. amaranthicolor
Loc.: C & E TX; not uncommon in Travis Co.
Form: herb; annual.
Notes – In my opinion, the foliage furnishes a food equally as good as *Chenopodium album*. The two species are very similar, with *C. album* being distinguished by a powdery white color on its leaf bases, whereas this color is replaced with pink on *C. giganteum*. There are conflicting data about the synonymy, introduced status, and range of both *C. album* and *C. giganteum* according to WFO, BONAP, USDA, POWO, and other plant databases. *C. giganteum* is likely introduced (WFO, POWO, Tropicos) and present in Texas (Tropicos, GBIF). Its classification as a variety of *C. album* (USDA) is incorrect (POWO, Tropicos, WFO), and this error may have contributed to some of these data conflicts.

Other *Chenopodium* spp. in Travis Co.:
Chenopodium berlandieri Moq.
Pitseed goosefoot
Loc.: all TX; rare in Travis Co.
Form: herb; annual.

Chenopodium pratericola Rydb.
Desert goosefoot
= Chenopodium desiccatum var. leptophylloides
Loc.: all TX except far E; rare in Travis Co.
Form: herb; annual.
Notes – *C. desiccatum* may also be in Central TX and has a perianth covering the fruit at maturity, whereas the perianth of *C. pratericola* spreads away. Both have linear leaves, as does *C. leptophyllum*, which has somewhat fleshy leaves. Other *Chenopodium* spp. in Central TX have wider and more toothed leaves.

Chenopodium album. Base photo: large plant in late August. Overlay, clockwise from top left: foliage (note whitish dusting of inner part of leaves); stems; range map (GBIF).

Left: *Chenopodium berlandieri* plant flowering in late April.
Right: *Chenopodium giganteum* plants in late April.
Range maps: GBIF.

APOCYNACEAE – Dogbane Family

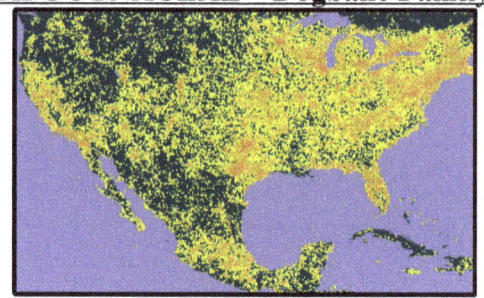

Asclepias distribution. Source: GBIF.

Asclepias spp.
Milkweed

Cahuilla: *kivat / kiyal*, Diné: *tc'il'apé'* – "plant, milk," Gosiute: *wañ'-go*, Hocąk: *maⁿhiⁿtc*, Kiowa: *zaip-ya-daw*
Loc.: all TX; very common in Travis Co. (9 spp.); 36 spp. in TX.
Form: herb, low shrub, shrub; perennial.

History

Despite their toxicity, at least 16 milkweed species were eaten by at least 31 Indigenous cultures. In most cases, only the young plant parts were used, and these were cooked before consumption.

Sometimes the plant tops (young shoots, fruits, and flowers) were cut and dried for winter storage (Kiowa, Menominee). *Asclepias speciosa* was the most widely used for food, with all parts except the roots eaten by Native peoples. *A. syriaca* was also commonly eaten, but its seeds were not eaten and its sap was not used for gum.

The gathering season for the shoots, flowers, and fruits of various milkweeds ranged from early to late spring. The seeds were harvested after June (Cahuilla).

Flowers – The flowers and buds of various milkweed species were eaten by the Northern Cheyenne, Crow, Diné, Haudenosaunee, Hocąk, Kiowa, Lakota, Menominee, Meskwaki, Ojibwe, Omaha, Pawnee, Ponca, Forest Potawatomi, Sioux, Winnebago, and Yokia. Species eaten include *Asclepias exaltata*, *A. incarnata*, *A. fascicularis*, *A. speciosa*, *A. syriaca*, and *A. tuberosa*.

They were eaten boiled (Northern Cheyenne, Crow, Haudenosaunee, Hocąk, Kiowa, Lakota, Meskwaki, Ojibwe, Omaha, Pawnee, Ponca, Forest Potawatomi, Winnebago), cooked in soups or stews (Northern Cheyenne, Hocąk, Menominee, Meskwaki, Ojibwe), cooked in grain meal mush (Menominee, Lakota), or raw (Yokia).

Fruits – The young fruits of various milkweed species were eaten by the Northern Cheyenne, Diné, Haudenosaunee, Hocąk, Kiowa, Meskwaki,

Ojibwe, Omaha, Pawnee, Ponca, Forest Potawatomi, Sioux, Tewa, and Winnebago. Species eaten include *Asclepias speciosa*, *A. syriaca*, and *A. tuberosa*.

The young fruits were eaten boiled (Northern Cheyenne, Crow, Haudenosaunee, Hocąk, Kiowa, Lakota, Meskwaki, Ojibwe, Omaha, Pawnee, Ponca, Forest Potawatomi, Winnebago, Sioux). They were dried for winter storage (Kiowa).

The green, immature fruits of *A. speciosa* were peeled, and the layer between the skin and seeds was eaten raw (Northern Cheyenne).

Seeds – The seeds of several species were eaten by the Cahuilla, Crow, and Jemez. Species eaten include *Asclepias speciosa* and possibly *A. eriocarpa*, *A. erosa*, or *A. fascicularis* (Cahuilla).

The seeds were ground into a meal and eaten (Cahuilla), or eaten raw when immature (Crow, Jemez).

Shoots – The young shoots of various milkweed species were eaten by the Cahuilla, Northern Cheyenne, Diné, Haudenosaunee, Hocąk, Hopi, Kawaiisu, Kiowa, Meskwaki, Myaamia, Ojibwe, Omaha, Pawnee, Ponca, Forest Potawatomi, Winnebago, Zuñi, and First Nations. Species eaten include *Asclepias erosa*, *A. incarnata*, *A. speciosa*, *A. syriaca*, *A. tuberosa*, *A. verticillata*, and possibly *A. eriocarpa* or *A. fascicularis* (Cahuilla).

The Myaamia only harvested early spring shoots less than ten inches tall and with more than four leaves. These shoots were boiled, with the water being changed more than once.

Leaves were parboiled (Cahuilla), boiled (Northern Cheyenne, Crow, Haudenosaunee, Hocąk, Kiowa, Lakota, Meskwaki, Ojibwe, Omaha, Pawnee, Ponca, Forest Potawatomi, Winnebago), roasted under hot ashes (Kawaiisu), cooked in soup or stew (Hopi, Menominee), or eaten raw (Northern Cheyenne).

Sap – The milky latex sap of various milkweed species was used as chewing gum by the Apache, Cahuilla, Northern Cheyenne, Flathead, Gosiute, Karuk, Kawaiisu, Kiowa, Shasta, Yurok. Species eaten include *Asclepias asperula*, *A. cordifolia*, *A. erosa*, and *A. speciosa*.

The stem was broken or cut, causing the outflow of latex, which was collected in a small vessel and dried overnight until solid (Cahuilla, Yurok), or was boiled to condense it (Cahuilla, Karuk, Kawaiisu). The gum was initially bitter, but after chewing and spitting, it became sweeter, and thereafter the juice was swallowed until the gum was discarded (Kawaiisu). The sap was also squeezed from the leaves and stems onto a heavy black clay to form gum (Apache).

Roots – The roots of several milkweed species were eaten by the Blackfeet, Diné, Haudenosaunee, Sioux, Tewa, and other American Indians. Species eaten include *Asclepias syriaca*, *A. tuberosa*, and *A. viridiflora*.

The roots were boiled (Haudenosaunee, Sioux, American Indians), eaten raw, or cooked in soup (Blackfeet). The roots were gathered in spring and were dried and stored for winter (Blackfeet).

Notes

WARNING – Milkweed is poisonous, and consuming it in any form can be harmful or fatal.

Asclepias species contain dangerous levels of cardiac glycosides, which can harm human health and potentially cause cardiac arrest.[25] The sap can also induce contact dermatitis.

In *A. syriaca*, the mature stems have the highest concentration of cardiac glycosides (about 2.6 mg/g digitoxin equivalent), while the roots and tops have the least (about 0.8 mg/g).[25] Cooking the shoots in the Myaamia method reduces the glycoside content by roughly half, but the ingestion of 100 grams of the cooked shoots would be comparable to 10 milligrams of digitoxin, an amount that can cause cardiac arrest.[25] No cases of ill effects as a result of eating milkweed have been reported among the Myaamia.[25]

In my personal experiments with eating milkweed, I have boiled the young top parts (flowers, fruits, and shoots) and discarded the water. I have experienced no ill effects from trying this with *A. asperula* and *A. viridis* on various occasions. I found the boiled young top parts to taste good. However, I do not recommend attempting this.

The fresh sap of *A. asperula* is very bitter. Once dried, it becomes less bitter and can be used as a chewing gum. The name "milkweed" comes from the milky white sap (latex) found in all *Asclepias* species.

Specific harvesting practices and cooking methods reduce the levels of these toxic compounds, but may not render milkweeds safe to eat. The historical methods described here may be incomplete or contain errors. Different milkweed species and their parts contain varying amounts of toxins. Individuals may vary in sensitivity to the toxins, and certain conditions can also exacerbate the effects. For these reasons, I do not recommend eating milkweed.

Asclepias asperula (Decne.) Woodson
Spider milkweed
= Asclepias capricornu, A. decumbens, Asclepiodora decumbens
Antelope horns, antelopehorn milkweed
Diné: *tjatíltee'íh* – "antelope horn," Gosiute: *pi'-wa-nûp*
Loc.: all TX except far E & S; common in Travis Co.
Form: herb; perennial.
Flowers: Mar-Oct (white, green).
Food: Gosiute (sap).

Asclepias tuberosa L.
Butterfly milkweed
= Acerates floridana
Butterfly weed, butterfly root, pleurisy root
Cherokee: *gu·gú* – "bottle / chigger" / *giga^dzuyaí* / *oneskwɔhí* / "chigger weed," Hocąk: *mąkąska* – "medicine white," Menominee: *kinokwe waxtsêtau* – "lying Indian or deceiver, and man-in-the-ground," Meskwaki: *atîste'i* / *päshtä'wûk* – "knob on roots," Omaha-Ponca: *maka^n saka* – "raw medicine" / *kiu maka^n* – "wound medicine"
Loc.: E, C, SE, N, & W TX; not uncommon in Travis Co.
Form: herb; perennial.
Flowers: May-Sept (orange, yellow).
Food: Sioux (flowers, fruits), First Nations (shoots), American Indians (roots).

Asclepias verticillata L.
Eastern whorled milkweed
= Asclepias galioides
Hopi: *piíñá* (Hough 1897:42, Hough 1898:148) / *piyüña* ~ "milk charm" / *pi:'ŋga*, Lakota: *waxpe tiŋpsila* – "turnip leaf," Zuñi: *ha'watseki* – "leaf, boy"
Loc.: E, SE, & C TX; uncommon in Travis Co.
Form: herb; perennial.
Food: Zuñi (shoots).

Asclepias viridiflora Raf.
Green comet milkweed
= Acerates viridiflora
Green antelopehorn milkweed
Lakota: *huciṅška* – "spoon plant"
Loc.: all TX (sparse S & W); not uncommon in Travis Co.
Form: herb; perennial.
Food: Blackfeet (roots).

Other *Asclepias* spp. in Travis Co.:
Asclepias oenotheroides Cham. & Schltdl.
Zizotes milkweed
Loc: all TX except far E; very common in Travis Co.

Asclepias texana A. Heller
Texas milkweed
Loc.: C & W TX; common in Travis Co.

Asclepias viridis Walter
Green antelopehorn
Loc.: E, NE, SE & C TX; common in Travis Co.

Asclepias. Clockwise from top left: *A. asperula* flowering in mid-April; *A. oenotheroides* with young, green fruit pods in mid-August; *A. texana* flowering in early August; *A. viridis* flower buds in mid-April with a close-up of opened flowers in late May. Range maps: GBIF.

ASTERACEAE – Composite Family

Liatris punctata Hook.
Dotted blazing star

= Lacinaria punctata
Plains gayfeather, dotted gayfeather, densespike blazing star, button snakeroot
Blackfeet: *mais-to-nata* – "crow root," Comanche: *ataβitsənoi*, Kiowa: *h-koñ-a*, Lakota: *tatečaŋnuğa* – "lumpy carcass," Meskwaki: *nîpinûskwɑ^h* – "summer weed," Tewa: *pi̱nnwi̱ki* – "mountain slope"
Loc.: all TX; common in Travis Co.
Form: herb, up to 3 ft. tall; perennial.
Flowers: Aug-Nov (pink, purple).

History

Roots – The corms were eaten by the Blackfeet, Comanche, Kiowa, and Tewa. They were eaten raw (Blackfeet, Comanche) or baked over a fire (Kiowa). Fresh corms were sometimes chewed, with the juice swallowed and the fibers spat out (Comanche). The corms were an important part of the Kiowa diet and were only gathered in the spring, as later in the season the roots had a "greasy taste" and were not eaten (Kiowa). The Blackfeet name refers to it being eaten by crows and ravens in the fall.

Notes

Character – This plant grows in prairie and loess hill habitats in Texas. In early spring, dotted blazing star is a rather inconspicuous, almost grass-like plant with a central stem covered in linear leaves. But its tall summer inflorescences bursting with vibrant, tassel-like lilac flowers make it easy to find, and it usually grows in patches.

Season – The corms are best harvested from late March through late April and become inedible by early summer.

Practice – These plants are most easily scouted when flowering, but their corms are not edible at that time. I note the locations of large patches by their flowers, then return in early spring, when the plants are growing new leaves and the stems are less than a foot high. These stems lead to the shallow corms, which can be dug out with a pointed stick.

I brush and wash off the dirt, then remove the dry, fibrous outer layer of the corms. I use a thin flint flake (or knife) to cut off the stem and roots, then cut into the bark and slide the flake under it, prying it off to reveal the smooth, wet inner layer.

The larger corms are not necessarily better than the smaller ones. At the center of each corm is a pecan-sized, fleshy tissue that is sweet, juicy, and not fibrous. It can be chewed up and eaten whole and raw. Surrounding this central tissue is a juicy, fibrous layer that varies in thickness; larger corms have more of this fibrous tissue, but the core tissue remains the same size. The fibrous outer layer is also sweet and can be chewed up, the juices sucked out, and the fiber discarded.

Cooking the corms gives them a soft, smooth, malt-like taste and makes them easier to chew. The fresh corms taste sweet, juicy, delicate, and mildly herbaceous, though the outer layer can be somewhat difficult to chew. Eaten raw, they are reminiscent of chewing a whole section of sugarcane, with a very sweet juice being sucked from the fibrous tissue.

Once the plants start getting larger, they become less sweet and juicy. By the time they flower and into the fall, they are very hard, fibrous, dry, darker in color, and not good to eat.

Additional notes:

Another species in the Asteraceae (composite family) worth mentioning is *Artemisia ludoviciana* (white sagebrush). It is found in all TX and is common in Travis Co. Its leaves were used for flavoring meat (Chiricahua and Mescalero Apache), making tea (Comanche), or chewing for flavor (Blackfeet). The plant had many medicinal uses, especially as an infusion or decoction, which was drunk for various ailments such as colds (Northern Cheyenne), sore throat (Kiowa, Meskwaki), and stomach problems (Dakota, Mexican Kickapoo, Kiowa, Omaha, Pawnee, Ponca, Winnebago). It was also used in poultices and as a favored incense for sweatbaths, purification, and ceremonial use (Arapaho, Northern Cheyenne, Comanche, Dakota, Kiowa, Omaha, Pawnee, Ponca, Winnebago). It has a strong, pleasant aroma and a sweet, medicinal taste.

Artemisia ludoviciana. L: patch of young plants in early April. R: single plant in late May (note greyish-green, soft foliage).

Liatris punctata. Base photo: flowering plant in mid-October. Overlay, clockwise from top left: young plant in late April (ready to harvest); close-up of stem and foliage in late April; close-up of flowers; large harvested corm, cut in half, in late April (note that the white center is not fibrous and the dark, layered outer part is, but both parts are sweet); range map (GBIF).

LAMIACEAE – Mint Family

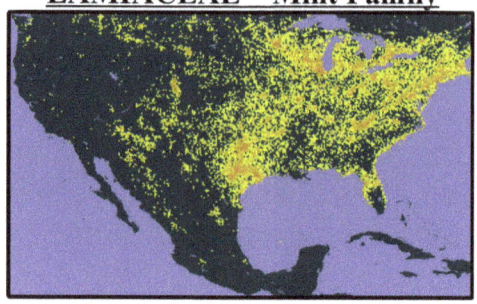

Monarda distribution. Source: GBIF.
Monarda spp.
Beebalm

Cherokee: *gɤˡshagí / gonsadí* – "it smells"
Loc.: all TX; very common in Travis Co. (4 spp.); 12 spp. in TX.
Form: herb, low shrub; annual, biennial, perennial.

History

Aboveground parts – The foliage, leaves, flowers, and stems of *Monarda* species were used for flavoring, tea, or internal medicine by the Apache, Cherokee, Creek, Dakota, Diné, Flathead, Haudenosaunee, Hopi, Isleta, Kiowa, Lakota, Menominee, Meskwaki, Ojibwe, Osage, Tewa, and Winnebago. Species used include *M. citriodora*, *M. fistulosa*, and *M. pectinata*.

The plants were used to flavor cooking (Apache, Hopi, Isleta, Tewa). Foliage was brewed into a pleasant, flavorful tea (Apache, Haudenosaunee). Leaves were dried and powdered as a spice (Flathead) or chewed raw for their flavor (Lakota). The plants were dried in bundles for winter storage, improving the aroma where they hung indoors (Diné, Flathead, Hopi).

The plants were gathered wild or cultivated (Hopi) and used for many medicinal purposes (Cherokee, Dakota, Flathead, Hocąk, Kiowa, Lakota, Menominee, Meskwaki, Osage, Tewa) as well as for their aromatic qualities (Dakota, Diné, Flathead, Kiowa, Omaha, Osage, Ponca).

Notes

Character – This genus can be found in most open grassland and prairie habitats in Austin, with *M. citriodora* being by far the most common *Monarda* species. The plants are attractive to pollinators as well as humans, with a unique look and stacked flower structure. They emit a pungent fragrance, reminiscent of citrus, thyme, and oregano.

Season – The plants begin growing in early spring, and can be collected for use as early as mid-spring. They continue maturing, with peak flowering from late spring to early summer, which is the optimal time for harvest. The last flowers dry out by mid- to late summer, when the plants in general begin to dry out and become unusable.

Nutrition – The potent odor of *M. citriodora* comes from its essential oils, with thymol accounting for 71% of the oils, and p-cymene (11%) and carvacrol (6%) making up the other dominant components.[22] These latter two oils are the main components of oregano (*Origanum vulgare*), while thymol and p-cymene dominate in thyme,[22] explaining their similarity to *Monarda* in odor and taste. Spanish thyme (*Thymus zygis*) only has 48% thymol,[22] perhaps making *M. citriodora* one and a half times more thyme-flavored than thyme. Oregano and thyme are also in the mint family (Lamiaceae). *M. citriodora* is a strong antioxidant and likely acts as a preservative and antibacterial agent.[22] The spice-like flavor and preservative properties of *Monarda* species help explain their historical use in foods.

Practice – I prefer to gather the plants during peak flowering from late spring to early summer, when they are fresh and bursting with flavor, though I also gather them throughout their availability. I break the base of whole plants, which usually consists of one main stem, and bundle them together in large bunches. I hang these indoors to dry, which also introduces a wonderful aroma. I use the plants for brewing tea, making kombucha, and as a substitute for Italian seasoning in various recipes. I only use *M. citriodora*, as it is the most common in my area, though the other species have a similar taste and smell.

Monarda citriodora Cerv. ex Lag.
Lemon beebalm

Lemon mint, wood betony, *betónica*
Hopi: *nanákopsi / nanakopsi* ~ "stacked flowers"
Loc.: all TX; very common in Travis Co.
Form: herb; annual, biennial, perennial.
Food: Hopi (flavoring in jackrabbit stew).

Monarda fistulosa L.
Wild bergamot

= Monarda comata, M. menthaefolia, M. menthifolia, M. stricta
Horsemint, Oswego-tea, wildbergamot beebalm, beebalm pennyroyal
Apache: *tołdai*, Cayuga: *ganu'da'*, Cherokee: *gows'agí* – "it smells," Dakota: *heȟaka ta pezhuta* – "elk medicine" / *wahpe washtemna* – "fragrant leaves," Diné: *'azee'ntoot'iijíh* – "medicine, [flowers in] whorls" / *t'óhnłtchiin* – "fragrant," Flathead: *tituw̓i*, Gosiute: *sutsigi'ⁿ*, Hocąk: *poaxų* – "sweat," Hopi: *nana'kofsi*, Kiowa: *po-et-oñ-sai-on* – "perfume plant," Lakota: *hexaka tapežuta* –

"elk medicine" / *hexaka tawote* – "elk food" / *waxpe waštemna* – "odorous leaves," Menominee: *oia'tcia näsikun* – "sneezing spasmodically," Meskwaki: *menaskwa'kûkį̂ʰ meskwanakį̂ʰ* – "smelling and red berries," Ojibwe: *weca' wûs wackwī' nek* – "yellow light," Omaha-Ponca: *pezhe pa* – "bitter herb" / *izna-kithe-iga hi*, Osage: *nidsida*, Pawnee: *tsusahtu* – "ill-smelling" / *tsostu*
Loc.: NE, E, & sparse C, S & W TX; uncommon in Travis Co.
Form: herb, low shrub; perennial.
Flowers: May-Oct (white, pink, purple).
Food: Apache, Flathead, Haudenosaunee, Hopi, Isleta, Lakota, Tewa.

Monarda punctata L.
Spotted beebalm

Dotted monarda, horsemint, wild bergamot, lemon mint
Diné: *'azee'ntoot'iijíh* – "medicine, [flowers in] whorls"
Loc.: all TX except far W; common in Travis Co.
Form: herb, low shrub; annual, biennial, perennial.
Food: no records (medicinal and material use only).

Other *Monarda* spp. in Travis Co.:
Monarda stanfieldii Small
Stanfield's beebalm
Loc: sparse C TX; uncommon in Travis Co.

Monarda citriodora. Base photo: flowering plant in late April (R) and late May (L). Overlay: stem close-up; range map (GBIF).

OXALIDACEAE – Woodsorrel Family

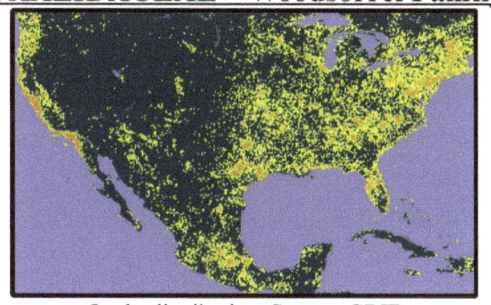 

Oxalis distribution. Source: GBIF.

Oxalis spp.
Woodsorrel

Cherokee: d͡ʒuni̇ᵈjɔsti / d͡ʒud͡zɔ́ isti / d͡ʒunazoyusti / d͡ʒuyasti – "sour" / gand͡ʒuʔdí, Haudenosaunee: deyu'yu'djis – "sour"
Loc.: all TX; very common in Travis Co. (8 spp.); 17 spp. in TX (5 introduced).
Form: herb, low shrub; annual, perennial.

History

Whole plant – The leaves, stems, flowers, fruits, and tubers of various *Oxalis* species were eaten raw or cooked by the Chiricahua and Mescalero Apache, Haudenosaunee, Karuk, Kiowa, Meskwaki, Omaha, Pawnee, Kashaya Pomo, Ponca, Forest Potawatomi, Tolowa, and Yurok. Species eaten include *O. corniculata*, *O. montana*, *O. oregana*, *O. stricta*, and *O. violacea*.

The leaves were chewed on long walks when sweating to relieve thirst and replace electrolytes (Kiowa) or to improve saliva flow (Cherokee). The plants were also eaten to stimulate appetite (Karuk). The leaves were combined with other greens to be cooked (Apache), eaten raw with dried fish (Tolowa, Yurok), or sweetened to make a dessert (Forest Potawatomi).

Notes – The tubers were pounded and fed to horses to make them fast (Pawnee, Omaha, and Ponca). Many Indigenous names for various *Oxalis* species translate into something like "sour herb." For example, (*O. violacea*) Apache: *itadnkodje* – "sour weed," Omaha-Ponca: *hade-sath* – "sour herb," Pawnee: *skidadihorit* – "sour like salt," Tewa: *'ogohep'e'ñœḇì* – "sour weed," or (*O. stricta*) Kiowa: *aw-tawt-añ-ya* – "salt weed," Hocąk: *xqwisku* – "weed sweet," Meskwaki: *wîskopi'pakwį̜ʰ* – "sweet weed" / *wîskapakûposi*.

The sialagogic (saliva-stimulating) effect of chewing woodsorrel, caused by its sour taste, is pronounced and was likely why the Kiowa considered them thirst-relieving, similar to the old survival trick of sucking on a small pebble. In Gaelic, one name for *O. acetosella* is *"greim saighdear*" – "soldier's mouthful," possibly in reference to a similar historical use of this plant.[39]

<u>Notes</u>

Character – This genus is one of the easiest to identify. It has compound leaves with three leaflets, forming a radially symmetrical trifoliate leaf. Each leaflet is distinctly heart-shaped and bilaterally symmetrical. They are all small herbs, usually only a few inches tall and rarely exceeding a foot in height. Although many people call them "clover," true clovers are in the genus *Trifolium* in the Fabaceae (legume family). True clovers have similar leaf forms but lack the heart-shaped leaflets and do not have a sour taste, though they are also edible. *Medicago polymorpha* (burclover) is another common, edible, clover-like herb in the area, resembling *Trifolium* but bearing spiny burs.

Woodsorrel species sometimes develop tubers, or small enlarged growths at the root tips where carbohydrates are stored. Most species are so tiny that these tubers are pea-sized at best. However, *O. tuberosa* in New Zealand produces tubers up to six inches long that are cultivated for their high carbohydrate content (83%).[36]

Season – Woodsorrel can be found year-round, but is most abundant and best harvested from late winter through spring.

Nutrition – The leaves of *O. pes-caprae* contain oxalic acid (98 g/kg dw) and vitamin C (3 mg/100 g dw).[17] The most concentrated macronutrients are 19% protein in the leaves, 73% carbohydrate in the stems, and 7% fat in the flowers, with lesser amounts of each of these macronutrients in the other tissues.[17] Compared to cultivated leafy vegetables, *O. acetosella* has high antioxidant capacity and is rich in carotene, vitamin C, tocopherols and xanthophylls.[58] Its oxalate content is high (about 125 mg/g dw), similar to rhubarb and spinach.[58]

Oxalis has a relatively high oxalate content for a leafy green, so precautions or proper preparation are advised, especially for individuals at risk of kidney stones.

Practice – This is probably the first wild edible plant I ever ate, discovered while waiting at my kindergarten bus stop. I have been eating it continuously for over three decades, and it is always pleasant.

It has a strong sour flavor, with mild sweet and herbal notes, and a crisp, tender texture. The young fruits, not mentioned above, taste similar and are best when green and immature, before the seed pods begin to dry out. The whole plant is edible, including the roots. The upper portions of the roots can be somewhat fibrous, but the lower roots often include tiny tubers, which are enjoyable to eat.

I use the aboveground parts in salads, eating them raw or cooking them like any other leafy vegetable. For storage, I simply dry the foliage and then use it in smoothies or for cooking.

Oxalis dillenii Jacq.
Slender yellow woodsorrel

= Oxalis lyonii
Loc.: all TX except W; very common in Travis Co.
Form: herb; perennial.
Food: no records.

Oxalis drummondii A. Gray
Drummond's woodsorrel

Loc.: all TX except far N; very common in Travis Co.
Form: herb; perennial.
Food: no records.

Other <u>Oxalis</u> spp. in Travis Co.:
Oxalis articulata Savigny
Pink-sorrel
Loc.: common in Travis Co.; introduced.

Oxalis corniculata L.
Creeping woodsorrel, 'ihi
Menominee: *wasa'wûs*, Onondaga: *deyuhiyu''djis awɛnu''gää'* – "sour plant"
Loc.: C, E, & S TX; common in Travis Co.; introduced.
Form: herb; perennial.

Oxalis debilis Kunth
Pink woodsorrel
Loc.: E, SE, & sparse C TX; not uncommon in Travis Co.; introduced.

Oxalis dehradunensis Raizada
West Indian woodsorrel
= Oxalis intermedia
Loc.: uncommon in Travis Co.; introduced.

Oxalis triangularis A.St.-Hil.
False shamrock
= Oxalis regnellii
Loc.: uncommon (cultivated) in Travis Co.; introduced.

Oxalis violacea L.
Violet woodsorrel
= Ionoxalis violacea
Apache: *itadnkodje* – "sour weed," Omaha-Ponca: *ḣade-sath* – "sour herb," Pawnee: *skidadihorit* – "sour like salt,," Tewa: *'ogohep'e'ñæbì* – "sour weed"
Loc.: E & C TX; rare in Travis Co.

Oxalis. Base photo: *O. drummondii* plants in late December. Overlay, clockwise from top left: *O. dillenii* range map (GBIF); flower in late April; fruiting plant in late April; *O. drummondii* range map (GBIF).

PLANTAGINACEAE – Plantain Family

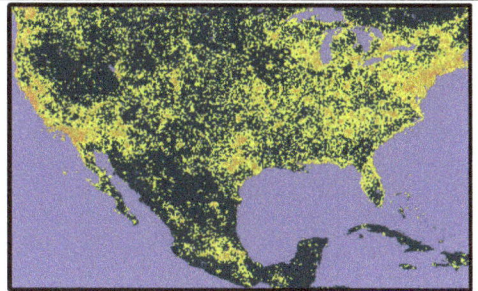

Plantago distribution. Source: GBIF.

Plantago spp.
Plantain

Cherokee: *nanɔudedɔˀtí* / *d͡ʒuyatałí talawadé·istí* – "grows around white oak tree"
Loc.: all TX; very common in Travis Co.; 14 spp. in TX (2 introduced).
Form: herb, low shrub, shrub; annual, biennial, perennial.

History

Seeds – Plantain seeds were eaten by the Pima, Seri, and Tohono O'odham. Species eaten include *P. ovata* and *P. patagonica*.

The seeds of desert Indianwheat (*P. ovata*) were an important food source for the Seri. Whole plants with mature seeds were pulled up and piled together. The seed spikes were rolled in the hands to release the seeds onto a fabric collection surface. Handfuls of the fallen seeds and chaff were then winnowed. The whole seeds were mixed with water and soaked for about half an hour. They formed into a glutinous mass, which was eaten without cooking, often with sugar added. More water was often added to make a beverage. This preparation was consumed both as food and as a remedy for stomachaches.

The Pima threshed and winnowed the dried seed heads of *P. ovata* and *P. patagonica*, then placed the seeds in water to make a mucilaginous drink.

The Tohono O'odham mixed a half cup of ripe *P. patagonica* seeds with a half cup of water and allowed it to stand for a short time before drinking it to treat diarrhea.

Notes

Character – Plantains are entirely edible, with young leaves that are generally palatable. The seeds grow on a central spike (raceme), which bears dozens of small seeds.

Season – The foliage of plantains is best eaten when young and tender, typically from late winter to late spring. The seeds ripen from early summer through fall.

Nutrition – One of the most popular commercial fiber supplements is "psyllium," which is the seed husks of *Plantago* species, including *P. ovata* and *P. psyllium*.

Practice – *Plantago* seeds can be gathered by removing the mature seed spikes, drying them, threshing out the seeds, and winnowing off the chaff. The seeds can then be soaked in water to form a mucilaginous drink, similar to chia.

The foliage of *Plantago* species can be eaten, and has a mild, herbaceous taste. As the leaves age, they become thicker, more pubescent, and more fibrous, but remain edible. They are best when the rosettes have just emerged and only have a few leaves. Once the plants begin growing their flowering stalks, I no longer harvest the foliage for food.

Plantago rhodosperma Decne.
Redseed plantain

Loc.: all TX; common in Travis Co.
Form: herb; annual.
Food: no records.

Plantago wrightiana Decne.
Wright's plantain

Loc.: all TX except far S; uncommon in Travis Co.
Form: herb; annual.
Food: no records.

Additional notes:

Plantago leaves are popular for poulticing in modern herbalism and were also historically used in similar ways by Indigenous peoples. The Cherokee used *Plantago* leaves as poultices for rheumatism, headaches, and sores. The Tolowa and Yurok used them as poultices on cuts and boils. They steamed the leaves, bent back the petiole, and used it along with the attached veins to peel off a layer of lamina (leaf sheath), which was then applied as a poultice in two or three layers. The Flathead, Hocąk, Houma, Menominee, Meskwaki, Ojibwe, and Ponca also used *Plantago* leaves for poultices, especially on cuts and sores, to draw out thorns, splinters, or pus, or for burns, scalds, bee stings, and snake bites.

The Tohono O'odham used the seeds of *P. ovata* and *P. patagonica* to remove foreign irritants stuck in the eye. A few seeds were placed in the eye, where they clung to the irritant so that both could be easily removed.

Top: *Plantago rhodosperma*. Base photo: young plant in early March. Overlay: fruiting heads in late April.
Bottom: *Plantago wrightiana* flowering plant in early June (in Llano Co.).
Range maps: GBIF.

SOLANACEAE – Nightshade Family

Capsicum annuum var. *glabriusculum* (Dunal) Heiser & Pickersgill
Chiltepín

= Capsicum annuum var. aviculare, C. hispidum var. glabriusculum
Cayenne pepper, American bird pepper, bush pepper, *chili pequín, chilequipín, chilchipin, aji*
Aztec: *quauhchilli*, Seri: *coquée quizil* – "chiles, little"
Loc.: S, SE, & C TX; common in Travis Co.
Form: herb, low shrub; annual, perennial.

History

Fruits – These peppers were eaten as a condiment by the Karankawa, Mexican Kickapoo, Seri, folk in Northern Mexico, and likely others. They were also eaten by the Mexican Kickapoo to stop hunger pains.

Notes

Character – The branches of this plant have a characteristic "zig-zag" form, with stems branching off at each angle. These junctures often feature a purple spot. The stems are otherwise mostly green, and the leaves are simple and soft rather than rigid. The plants grow in moderately dry to very dry, thin soils.

This is actually the same species as common garden pepper, serrano, cayenne, bell pepper, and many other chiles, which are all different varieties of *Capsicum annuum*. The wild ancestor of these plants was domesticated in Mesoamerica around 6,000 years ago and probably resembled chiltepín.

The name "chiltepín" or "chiltecpin" is a Nahuatl word meaning "flea chili," referring to the small size of the fruits. Spanish speakers likely converted this into "chili pequín," also in reference to the small size. The commercial cultivar of this variety produces larger, more elongated fruits and is most commonly called "chili pequín," whereas the wild plants with smaller, more globular fruits are commonly called "chiltepín." Birds are not affected by capsaicin and eat and disperse these fruits, giving it the name "American bird pepper."

Capsaicin is the "hot" compound in peppers. It activates receptors that detect heat and abrasion, can trigger endorphin release, and may improve pain tolerance. It does not cause ulcers or damage taste buds.

Season – The ripe fruits peak in Austin from mid-September to mid-October, but can be found from late summer to late fall, and even sometimes persist on the plants until the following spring, albeit old and dried.

Nutrition – Dried cayenne peppers contain 12% protein, 17% fat, 57% carbohydrate, and 27% fiber, vitamin C (76 mg/100 g), potassium (2010 mg/100 g), and vitamin A (41600 IU).[71] They are excellent or extremely high sources of vitamin A, vitamin C, and potassium. Chiltepín scores 50,000-100,000 on the Scoville heat scale. In my personal experience as an avid spicy pepper enthusiast, the heat of wild chiltepíns is comparable to that of habaneros.

Practice – I like to gather the fruits by hand and eat them fresh or dry them for long-term storage. They are my favorite kind of chili pepper, and I have tried every variety I can obtain. Their flavor is similar to habanero, which is somewhat distinct from other peppers. The closest comparison is a small, globular pepper from the Peruvian Amazon that is a yellow-orange in color, called "ají charapita," which belongs to the species *Capsicum frutescens*.

In some years, I find chiltepín fruiting abundantly, while in other years they produce less. In the latter case, I keep my harvests minimal. Chili pequín makes an excellent garden plant, requiring almost no care, yet still fruiting prolifically, so I highly recommend planting it. Chili pequín is available in many grocery stores, and its seeds can be planted. I eat it, or habanero, daily.

Additional notes:

Among the Mexican Kickapoo, chiltepín fruits were crushed and rubbed on the back to relieve pain from pneumonia, rubbed on the nipple of a nursing mother to wean her child, and sprinkled on objects to prevent a child from chewing or eating them. *Capsicum baccatum* (bird pepper, *locato*) is found in far S TX, and its fruits were eaten by the Tohono O'odham as a condiment or spice.

The fruits of at least 7 *Physalis* spp. (groundcherries, Solanaceae), were historically eaten by at least 14 Indigenous peoples.

Solanum elaeagnifolium (silverleaf nightshade, Solanaceae) fruits were used by the Pima and Zuñi for curdling goat's milk. They were also "eaten, or boiled to make a syrup" by the Isleta Pueblo, though they were "thought to have a laxative effect" and are generally considered inedible. It is in all TX and very common in Travis Co. *S. triquetrum* (Texas nightshade) fruits were boiled by the Zuñi, ground in a mortar, mixed with chile and salt, and eaten as a condiment with mush or bread, though they are also not considered edible. It is in all TX except N and common in Travis Co.

Capsicum annuum var. *glabriusculum*. Base photo: fruiting plant in late October. Overlay, clockwise from top right: purple coloration on branch forks; close-up of ripe fruit in mid-September; fruiting branch (note zig-zag branching pattern); range map (GBIF).

URTICACEAE – Stinging Nettle Family

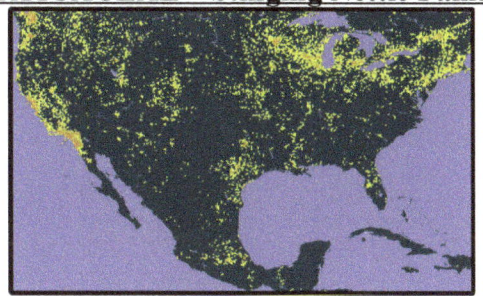 

Urtica distribution. Source: GBIF.

Urtica spp.
Nettle

Loc.: all TX (sparse N); common in Travis Co. (2 spp.); 4 spp. in TX (1 introduced).
Form: herb; annual, perennial.

History

Foliage – The leaves and stems of *U. gracilis* were eaten raw, boiled (Cahuilla), or cooked like spinach (Haudenosaunee).

Notes

Character – Stinging nettles are covered with hair-like structures that, when touched, can inject formic acid along with other compounds into the skin, similar to ants (Formicidae). These hairs must be stiff to puncture the skin and deliver their contents. Cooking the plants soften the hairs, preventing them from stinging and making them safe to eat. The new growth of leaves and stems at the top of the plant may have underdeveloped hairs incapable of stinging, and this young foliage can be eaten raw. The entire plant is edible when cooked.

U. gracilis is generally not found in Texas, but I have observed it growing wild in persistent, healthy patches in Austin in distantly separated locations. *U. chamaedryoides* is the common species in Austin and Texas, and is easily distinguished by its heart-shaped or broad and rounded leaves, in contrast to the long, narrow leaves of *U. gracilis*.

Season – In Austin, *U. chamaedryoides* can be found year-round but is mostly available from late winter through early spring. *U. gracilis* can be found up to midsummer, and possibly later.

Nutrition – With 30% protein by dry mass, *U. dioica* leaves have an exceptionally favorable amino acid profile compared to most green plants, along with 4% fat and 10% fiber.[11] One hundred grams of blanched stinging nettle provide 416% the daily value for vitamin K, 67% for vitamin A, 12% for vitamin B2, 37% for calcium, 34% for manganese, and 14% for magnesium.[11] They are an excellent or good source of all of these vitamins and minerals except magnesium and B2, for which they are a moderate source. Nettle also contains numerous other vitamins and minerals in significant amounts.[11] The seeds are similarly nutritious, with higher concentrations of fatty acids.[11] The plant has high antioxidant activity and is rich in polyphenols and carotenoids.[11] It is regarded as one of the healthiest foods in the world.[11]

Practice – I gather heartleaf nettle from late winter through spring. I usually wear leather gloves to cut the plant tops and place them in a bag. They can be dried for storage, which also neutralizes the stinging hairs. To prepare fresh or dried nettle, I parboil or blanch them, boiling for one to several minutes and sometimes immediately cooling them in ice water to improve texture, flavor, and nutrient retention.

It is possible to fold the leaves in such a way so they can be eaten raw. The lower surface of the leaves has fewer or smaller hairs, which allows careful manipulation in the fingers to fold the upper surface inward on itself. The folded leaf can then be eaten, destroying the hairs by mastication before they can sting. I sometimes do this to eat nettle in the field. I also occasionally eat the very youngest top parts raw, as they are not yet capable of stinging.

The seeds are borne on spikes at the tops of the plants. They are tiny but produced in abundance. I usually gather the fruiting plants with seeds and cook them whole along with the rest of the plant tops. It is possible to thresh out the seeds and sift them to use just the seeds for cooking, but I usually leave them in the plant.

Urtica chamaedryoides Pursh
Heartleaf nettle

Loc.: C, SE, S, SW, E & sparse N TX; common in Travis Co.
Form: herb; annual.
Flowers: Jan-Nov (white, pink, purple).
Food: no records.

Urtica gracilis Aiton
Stinging nettle

= Urtica lyallii, U. californica, U. dioica ssp. gracilis, U. holosericea, U. strigosissima, U. viridis
California nettle, western nettle
Cahuilla: *chikishlyam*, Cayuga: *gohe*ʻ*'cra's*, Cherokee: *tɔledá* / *toledaata^{d}sastí* – "stinging on you," Cree: *masān*, Flathead: *cćaxelshp* – "sting leaf," Gosiute: *tinʼ-ui-gop*, Karuk: *'akviin* – "bright" / *anievxaat* – "smells like under arm," Kawaiisu: *kʷitʃiʔatabi* / *kutsiʔatabi*, Klamath: *sleds*, Lakota: *čaŋíčaxpe* hu – "woody whip plant," Osage: *hadoga*, Menominee: *säʻnap*, Ojibwe: *masan* – "woods," Omaha-Ponca: *hanuga-hi* / *manazhiha-hi*, Forest Potawatomi: *masan* – "itching"
Loc.: sparse N TX; rare in Travis Co.
Form: herb; perennial.
Food: Cahuilla, Haudenosaunee.

Urtica. Base photo: *U. chamaedryoides* young plants in late December. Overlay, top: *U. chamaedryoides* range map (GBIF), and stinging hairs. Overlay, bottom: *U. gracilis* foliage in mid-April and range map (GBIF).

MONOCOT HERBS

AMARYLLIDACEAE – Amaryllis Family

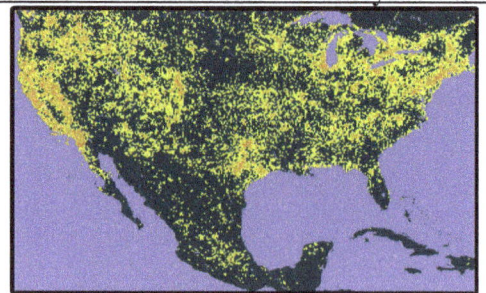

Allium distribution. Source: GBIF.

Allium spp.
Onion

Northern Cheyenne: *xaóe-hehestavo* – "skunk nuts" / *tóhtooʔe-xaóe-nėstavo* ~ "prairie skunk."
Comanche: *pakǿ:k* (large onions) / *tʔdiekǿ:k* (small onions), Diné: *tl'ohtchin* – "grass, smelly."
Lakota: *monžonxe* – "earth, to bury," Osage: *pšin*, Northern Ute: *wisi-sik^wu*
Loc.: all TX; common in Travis Co. (3 spp.); 17 spp. in TX (5 introduced).
Form: grass-like herb; perennial.
Notes – There are poisonous look-alikes that can easily be distinguished by the absence of an obvious onion smell to the crushed foliage.

History

Whole plant – Wild onions were eaten by the Chiricahua, Mescalero, and Western Apache, Blackfeet, Calpella, Cahuilla, Northern Cheyenne, Comanche, Concow, Dakota, Diné, Flathead, Gosiute, Haudenosaunee, Hopi, Isleta, Kawaiisu, Mexican Kickapoo, Little Lake, Menominee, Nomlaki, Ojibwe, Omaha, Osage, Pawnee, Pomo, Ponca, Forest Potawatomi, Seri, Tewa, Tohono O'odham, Northern Ute, Wailaki, Winnebago, Yokia, Yuki, Plains Natives, Natives of the Upper Mississippi Valey and Great Lakes, Great Basin Mojave Desert tribes, Natives in the Pacific Northwest, California Natives, Texas Natives 5,000 years ago and in the Archaic, folk in northern Mexico, Indigenous people in Tenochtitlan around 1519, and other Natives. Species eaten include *A. bisceptrum, A. brevistylum, A. canadense, A. cernuum, A. drummondii, A. geyeri, A. haematochiton, A. macropetalum, A. schoenoprasum, A. tricoccum, A. unifolium, A. validum*, and likely others.

  The plants, especially the bulbs, were eaten fresh and raw, cooked in soups and stews, used to flavor meat or other foods, or were dried for storage. They were also cooked in earth ovens, then were sun-dried (Great Lakes & Upper Mississippi Valley Natives, Texas Natives in the Archaic). The bulbs were rubbed in hot ashes to singe them and improve their taste, then were eaten or dried for winter use, being soaked before eating (Diné). The larger plants were braided together and roasted over a fire, while the smaller plants were boiled (Comanche). Produce from a variety of *Allium* species was sold in markets by Indigenous people in Tenochtitlan around 1519.

Notes

Character – All *Allium* species are likely edible for humans, and the genus is widespread throughout the Northern Hemisphere. They form an important part of the diet for peoples globally, most notably in the forms of *A. cepa* (garden onion) and *A. sativum* (garlic), which are cultivated worldwide.

Allium is easy to identify by its distinct odor of onions when the foliage is crushed. Toxic look-alikes lack this characteristic onion smell. Crow poison (*Nothoscordum bivalve*) looks very similar to *Allium* and grows in the same habitat. There are conflicting reports of its edibility or toxicity, but it can be distinguished by its absence of onion scent. Crow poison typically has from 2 to 10 flowers per stalk, while *Allium* usually has 8 or more, though sometimes only one or two. Unlike grasses, *Allium* leaves have no sharp edge and appear as flattened tubes in cross-section.

The whole plant, including the flowers and "fruits," is edible and flavorful. These "fruits" are actually sessile bulblets, which are small, globular structures borne at the tops of the fruiting stalks, below the true fruits, which are dry and open to reveal the seeds. These sessile bulblets drop off the plant when disturbed and can grow into new plants through vegetative reproduction. They are the best part of the mature plants, with a strong onion taste, crisp texture, and interesting form, excellent for topping salads or meat dishes.

A. canadensis and *A. drummondii* are the most common species in Austin. *A. neapolitanum* is an introduced species that also occurs in the area. *A. canadensis* is larger, has white flowers, and is more often found in moist soils, whereas *A. drummondii* is smaller, has white, pink, or purplish flowers, and is more often found in drier soils.

Season – As the plants mature and fruit, they become fibrous and tough. While they can be found year-round, they are preferable during their earlier growth stages, which sometimes occur in the fall and winter but usually are found from late winter through spring. In late winter, they appear grass-like in form and furnish excellent greens, similar to chive tops. The bulbs are best harvested in the spring, when they are full-sized, crisp, and flavorful. From late spring through summer, the sessile bulblets are the prime part to harvest.

Nutrition – Fresh wild onions in Colorado contain 2% protein, 0% fat, 21% carbohydrate, 6% fiber, calcium (438 mg/100 g), potassium (272 mg/100 g), magnesium (44 mg/100 g), copper (9 mg/100 g), other trace minerals, and likely contain substantial amounts of vitamin C.[43] Wild onions are an excellent nutritional complement to bison meat.[43] They are exceptional sources of copper, good sources of calcium, and moderate sources of magnesium.

Practice – I gather the foliage in late winter to early spring by tearing or cutting it, using it as a substitute for chives. In mid- to late spring, I harvest the bulbs by carefully grasping the plant at its base and gently pulling until the bulbs are uprooted. If it feels like the plant will break before the bulb will come up, I use a pointed stick to loosen up the surrounding soil, then pull up the bulbs. From late spring to midsummer, I typically only gather the sessile bulblets by rubbing them off the plants.

I use all parts of the wild onions just as I would regular onion, or dry them for storage. I dry the whole plants, coarsely grind them up, and store them in a sealed container.

Allium canadense L.
Meadow garlic

= Allium acetabulum, A. continuum, A. reticulatum, A. fraseri, A. hyacinthoides, A. arenicola, A. microscordion, A. movilense, A. mutabile, A. zenobine
Canada onion, Canadian meadow garlic, wild onion, wild garlic
Dakota: *pshi^n*, Omaha-Ponca: *ma'^{n}zho^{n}ka-mantanaha*, Pawnee: *osidiwa*, Winnebago: *shi^{n}hop*, Haudenosaunee: *gahadagonka'*, Menominee: *sikaku'sia*, Meskwaki: *shîkako'aki^h* – "skunk weed," Onondaga: *u'no̱'sa' ga'no̱suha'ha'*, Forest Potawatomi: *fîgaga'wûnj* – "skunk plant"
Loc.: all TX except W; common in Travis Co.
Form: grass-like herb, up 2 ft. tall; perennial.
Flowers: Mar-July (white, pink).
Notes – The Algonquin word "*shîkako*," from "*skîkak*" – "skunk" is the origin of the name of Chicago, and it is similar to the Menominee and Meskwaki words for this species and *Allium tricoccum*, which grow abundantly around Chicago. Chicago is first mentioned as "the portage of Chicagou" in 1682, and as "Chicagou" in 1687 during the La Salle Mississippi expedition (Weddle 1987). Members of the 1674 expedition to the site of Chicago ate a large amount of onions there (Smith 1933).
Food: Dakota, Haudenosaunee, Menominee, Omaha, Pawnee, Ponca, Forest Potawatomi, Winnebago.

Allium drummondii Regel
Drummond's onion

= Allium nuttallii
Cebolla del monte
Loc.: all TX; very common in Travis Co.
Form: grass-like herb; perennial.

Flowers: Mar-May (white, pink).
Food: Northern Cheyenne, Mexican Kickapoo, Texas Natives in the Archaic.

Other *Allium* spp. in Travis Co.:
Allium cepa L.
Garden onion, *cebolla*
Hopi: *siːʼwi / sioʔuʼyi* – "onion, to plant," Seri: *hehe ccon* – "plant that-reeks"
Loc.: introduced; cultivated.

Allium neopolitanum Cirillo
White garlic
Loc.: sparse C & E TX; common in Travis Co.; introduced.

Allium sativum L.
Garlic, *ajo*
Diné: *tłʼohtchintshoh* – "grass, smelly, big," Seri: *haaxo*
Loc.: introduced; cultivated.

Additional notes:

They are not related to *Allium*, but a couple more species are worthy of inclusion in this book: *Commelina erecta* and *Tinantia anomala*. I have found no evidence of historical use of either of these native plants as food by Indigenous peoples. However, the entire plants are edible and palatable. Both, along with *Tradescantia* spp. (listed in this section), are in the Commelinaceae (Commelina family), which includes many edible herbs. *Commelina erecta* is found in all TX, and is common in Travis Co. *Tinantia anomala* is found in C, SW, and coastal SE TX, and is common in Travis Co. The two species look and taste similar. They are crisp, mildly-flavored, and slightly mucilaginous. Unlike many edible herbs, the mature parts are as good as the young parts. They can be eaten fresh or dried for storage, and the flowers are an attractive garnish.

L: *Tinantia anomala*, R: *Commelina erecta*.

Allium canadense. Base photo: freshly dug plants in mid-February. Overlay, clockwise from top left: range map (GBIF); sessile bulblets on fruiting plant (seeds are glossy black) in mid-June; sessile bulblets on flowering plant in late April; harvested and cleaned plants in mid-February.

Allium drummondii. Base photo: flowering plants in late March. Overlay, clockwise from top left: range map (GBIF); flower close-ups, white, pink.

COMMELINACEAE – Commelina Family

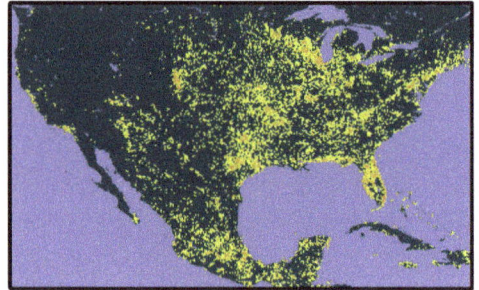

Tradescantia distribution. Source: GBIF.

Tradescantia spp.
Spiderwort

Loc.: all TX; common in Travis Co. (7 spp.); 17 spp. in TX (2 introduced, 1 hybrid).
Form: herb, low shrub, shrub; annual, perennial.

History

Aboveground parts – The foliage of *T. occidentalis* was cooked as greens by the Hopi and highly prized as a food. The aboveground parts of *T. virginiana* were gathered in the spring eaten as greens by the Cherokee.

Notes

Character – *Tradescantia* species readily hybridize and can be difficult to distinguish. *T. pallida* (purple heart) is the most distinct in Austin and is a common landscaping plant with broad leaves and foliage that is often purple. *T. gigantea* is the largest and most common in the area. All these species have edible foliage.

Season – The plants can be found from late winter through summer, but are especially abundant and flowering from late March to early April.

Nutrition – *T. fluminensis* foliage contains high phenolic content (5 to 12 mg GAE/g) and exhibits strong antioxidant activity.[32]

Practice – I like to gather *T. gigantea* foliage and aboveground parts to eat fresh and raw in the spring. It has a mild, herbaceous flavor with a crisp, slightly mucilaginous texture.

Tradescantia gigantea Rose
Giant spiderwort

Loc.: C & E TX; common in Travis Co.
Form: herb, up to 3 ft. tall; perennial.
Flowers: Mar-Apr (white, pink, blue, purple, violet).
Food: no records.

Tradescantia occidentalis (Britton) Smyth
Prairie spiderwort

Hopi: *pasomi* / *paso'mi*, Meskwaki: *mêkosike'shîkêkj^h* / *pûkwoskûk sakä'sêkûk*
Loc.: all TX; not uncommon in Travis Co.
Form: herb; perennial.
Flowers: June-July (pink, purple).
Food: Hopi.

Tradescantia ohiensis Raf.
Bluejacket

= Tradescantia reflexa
Ohio spiderwort
Lakota: *čaŋxloğaŋ paŋpaŋla* – "soft weed"
Loc.: all TX except far N and far S; uncommon in Travis Co.
Form: grass-like herb; perennial.
Flowers: Mar-Sept (white, pink, blue).
Food: no records.

Tradescantia virginiana L.
Virginia spiderwort

Cherokee: *tawaló* / *tagwa·lí* / *ta'gwa·lá* / *yonuǹigistí*
Loc.: sparse E, C, & SE TX; uncommon in Travis Co.
Form: grass-like herb; perennial.
Flowers: Mar-Aug (white, blue, purple).
Food: Cherokee.

Tradescantia. Base photos (L & R): *T. gigantea* flowering plants in mid- to late February (range map on bottom). Overlay: *T. occidentalis* flowering plant in late March. Range maps: GBIF.

WETLAND PLANTS

CYPERACEAE – Sedge Family

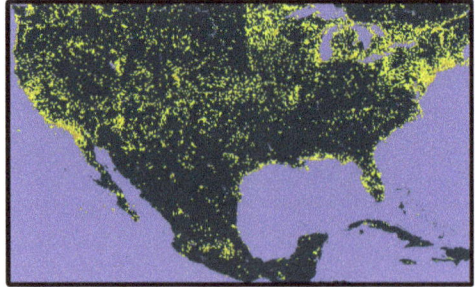

Schoenoplectus distribution. Source: GBIF.

Schoenoplectus spp.
Bulrush

Loc.: all TX; not uncommon in Travis Co. (3 spp.); 11 spp. in TX.
Form: grass-like herb, up to 12 ft. tall; perennial; water habitat.

History

Flowers – Bulrush pollen was eaten by the Chiricahua Apache, Cahuilla, and other Western Natives. Species eaten include *S. acutus*.

The flowering tops were cut off, dried, and then shaken over spread-out buckskin, as cloth would not hold such fine powder (Apache). The pollen was collected and mixed with meal or used plain to make cakes or mush (Cahuilla, Western Natives).

Seeds – Bulrush seeds were eaten by the Cahuilla, Klamath, Northern Paiute, and other Natives. Species eaten include *S. acutus* and *S. pungens*.

The seeds were beaten off the plants into baskets and eaten raw, mixed with other meal, ground into a mush, or made into cakes (Cahuilla, Western Natives). The fruiting stalks were also cut, gathered, and dried (Northern Paiute). These were then placed on a metate, and the seeds were worked out with a mano. The separated seeds were ground into meal and made into mush for consumption.

The seeds of a similar plant, *Bolboschoenus maritimus*, were also eaten by the Northern Paiute. When the seeds dropped into the water, they accumulated in drifts along the shores, from where they were gathered. They were parched by shaking them in baskets with live coals. The parched seeds were stored, ground into meal, and made into mush.

Shoots (culms) – A long piece of lateral root (rhizome) was dug from the mud by the Northern Paiute, and the new shoots emerging from it were broken off. The leaves were peeled from these shoots, and the underlying crisp, white tissue was eaten fresh and raw. Such young shoots were also eaten raw or cooked by other Western Natives, including the Northern Cheyenne, Cree, and Gosiute. Species eaten include *S. acutus*.

Shoots (basal swellings) – The small bulbous portion located where the root turns, or the basal portions of stalks, were eaten fresh and raw or cooked in soup (Chiricahua Apache, Cree, Dakota, Kawaiisu, Ojibwe, and Northern Paiute). Species eaten include *S. acutus* and *S. tabernaemontani*. These basal swellings were gathered in August (Northern Paiute).

Roots – Bulrush rhizomes were eaten by the Chiricahua Apache, Cahuilla, Cocopah, Cree, Gosiute, Haudenosaunee, Ojibwe, Northern Paiute, Western Natives, Natives of the Upper Missouri River, Natives in California, and other Natives. Species eaten include *S. acutus*.

The white, starchy, tuberous rootstocks (rhizomes) were dried and ground into a sweet meal used to make mush or cakes (Cahuilla, Haudenosaunee, Western Natives). Fresh rhizomes were also boiled, roasted, or merely peeled and chewed to extract the juices (Cocopah, Northern Paiute, other Natives). The juice from the bruised fresh rhizomes was eaten raw (Haudenosaunee). A sweet syrup was made by bruising the rhizomes, boiling them for several hours, and then pouring off the liquid (Western Natives, other Natives). When gathered in midsummer, the rhizomes had a sweetish taste and were eaten raw (Ojibwe). Rhizomes were gathered year-round but were preferred from early spring through summer (Apache, Ojibwe).

The small tubers of *Bolboschoenus maritimus*, which were a little bigger than a fingernail, were eaten by the Gila River Pima by pulling up the whole plant after it had flowered.

Notes

Character – These plants resemble cattails, but the leaves are round in cross-section, whereas cattail leaves are flattened. The flowers and fruits are also distinct, being borne on clusters that are spread out rather than a single erect spike like as in cattails.

They are found on the edges of perennial water sources such as rivers and lakes, where they form dense stands. *S. californicus* is the most common species in Austin. The various *Schoenoplectus* species are likely all usable for food in similar manners to those described above.

Caution should be exercised when harvesting aquatic or wetland plant in urban areas, as they readily absorb pollutants.

Season – The pollen is available from spring through summer. The emerging shoots are available in the spring. The basal swellings and

rhizomes can potentially be harvested year-round.
Nutrition – The rhizomes gathered by the Northern Paiute contained about 0.5-0.6 kcal/g, 1.5% protein, and 14% carbohydrate (Fowler 1990).
Practice – I have not tried harvesting this plant, as they are not widespread in the area. The largest populations in relatively unpolluted waters in Austin are accessible only by boat.

Schoenoplectus californicus (C.A.Mey.) Soják
California bulrush

= Elytrospermum californicum, Schoenoplectus californicus (C.A. Mey.) Palla
Loc.: all TX except NW & far N; not uncommon in Travis Co.
Form: grass-like herb, up to 12 ft. tall; perennial; water habitat.
Flowers: Apr-Aug (white, yellow).
Food: no records.

Schoenoplectus pungens (Vahl) Palla
Common threesquare

= Scirpus pungens
Sharp club-rush, chairmaker's-rush
Northern Paiute: midɨbui, Gila River Pima: *vak*
Loc: all TX; rare in Travis Co.
Form: grass-like herb, up to 6 ft. tall; perennial; water habitat.
Flowers: Mar-Aug (green, brown); fruits spring-summer.
Food: Northern Paiute.

Schoenoplectus tabernaemontani (C.C.Gmel.) Palla
Softstem bulrush

= Schoenoplectus validus, Scirpus glaucus, S. tabernaemontani, S. validus
Great bulrush
Dakota: *psa,* Omaha-Ponca: *sa-hi,* Pawnee: *sistat*
Loc.: all TX, rare in Travis Co..
Form: grass-like herb, up to 10 ft. tall; perennial; water habitat.
Flowers: Apr-May (red).
Food: Dakota.

Additional notes:

The stems of *Schoenoplectus* species were commonly used to weave mats and baskets (Cahuilla, Dakota, Karuk, Kawaiisu, Klamath, Menominee, Meskwaki, Ojibwe, Omaha, Osage, Pawnee, Ponca, Forest Potawatomi, other Natives), especially mats, which served for thatching, walls, flooring, sitting, trays, and more. The stems were placed parallel and bound together with cordage or strips of *Yucca* leaves.

Schoenoplectus californicus. Top: plants in a river marsh in late September. Bottom: plants bordering lake in late May. Overlay: close-up of top plants; range map (GBIF).

NELUMBONACEAE – Lotus Lily Family

Nelumbo lutea (Willd.) Pers.
American lotus

= Nelumbo pentapetala, Nelumbium luteum, N. pentapetalum, Nymphaea pentapetalum
Yellow lotus, yellow water lotus, water chinquapin, water chiquapin, yellow nelumbo
Comanche: *keʎiats*, Dakota: *tewape*, Meskwaki: *waki'pimînĵ^h* / (the seeds) "*oshkîshi'kûkĵ^h*," Ojibwe: *wâgipin* – "crooked root" / *wesawasa' kwune'k odîte'abûg* – "yellow light, flat heart-shaped leaf," Omaha-Ponca: *tethawe*, Oto / Quapaw: *tarowa / taluwa* – "hollow root," Pawnee: *tukawiu*, Forest Potawatomi: *wagipîn* – "crooked potato," Winnebago: *tserop*, Natives of the Great Lakes: *poke-koretch*
Loc.: E, SE, NE, sparse N & C, & far S TX; uncommon in Travis Co.; only *Nelumbo* sp. in TX.
Form: herb, up to 6 ft. tall; perennial; aquatic.
Flowers: June-Sept. (yellow).

History

Seeds – The large seeds of American lotus were eaten by the Dakota, Meskwaki, Ojibwe, Omaha, Osage, Pawnee, Ponca, Forest Potawatomi, Winnebago, tribes of Upper Mississippi Valley and Great Lakes, and other Natives.

The seeds were generally eaten raw or cooked. They were cracked and shelled and cooked as a soup or were roasted under hot coals. Roasted seeds could be ground into meal or mixed with corn.

Roots – The rhizomes were eaten by the Comanche, Dakota, Meskwaki, Ojibwe, Omaha, Osage, Pawnee, Ponca, Forest Potawatomi, Tonkawa, Winnebago, and other Natives.

The rhizomes were gathered by searching for them while wading with the toes, working the mud away by foot, and pulling them up with a hooked stick (Dakota, Omaha, Pawnee, Ponca, Winnebago). The Meskwaki and Ojibwe cut off the terminal shoots at either end of the underground rootstock and ate the remainder. The Forest Potawatomi gathered the two terminal shoots of an underground rootstock, which serve as starch reservoirs at the end of the growing season and are about the size and shape of a banana. The Omaha called bananas "*tethawe ega^n*" – "the things that look like *tethawe*," since the rootstocks of this species look like small bananas. The rhizomes were gathered in winter, with the plants located by their dried stems above the ice (Great Lakes Natives).

The rhizomes were cooked freshly harvested or after being dried for winter use. They were boiled (Comanche, Osage), roasted (Great Lakes Natives), or formed into cakes (Tonkawa). To preserve them, the rhizomes were peeled, cut into pieces about an inch long, strung on a cord, and dried, sometimes over smoke (Dakota, Ojibwe, Omaha, Meskwaki, Pawnee, Ponca, Forest Potawatomi, Winnebago, Great Lakes Natives).

The dried pieces were later soaked or boiled to prepare them for eating, or sometimes ground into a meal that was made into a cake (Tonkawa).

They were an important food source for the Dakota, Omaha, Pawnee, Ponca, Winnebago, and other peoples and were gathered in great abundance by the Tonkawa in Central Texas. The Meskwaki seasonally migrated to areas where they could harvest them in plenty. American Indians may have introduced and naturalized this species far beyond its native range to use its tubers and seeds for food.

Notes

Character – Water lilies (Nymphaeaceae) such as *Nuphar* and *Nymphaea* have division or "cut" in their leaves, whereas lotus lilies (Nelumbonaceae), such as *Nelumbo lutea*, have complete, circular leaves with no divisions. Despite their superficial similarity, the two families are distantly related. Though, *Nuphar* and *Nymphaea* were actually used for food similarly, with the seeds and rhizomes of species in both genera being historically eaten by Indigenous North Americans.

Nelumbo lutea is found throughout eastern North America, from Miami to Michigan and from the Atlantic coast to Kansas, but is not common in Central Texas. It grows in lakes, ponds, and slow-moving parts of rivers and creeks with muddy bottoms. Its huge flowers produce large fruits filled with nut-like seeds. Once shelled, the seeds can be eaten raw or cooked. The immature green fruits can be broken open, with their tender seeds eaten raw or cooked. The dried, stored seeds of *Nelumbo nucifera* are prone to moisture and mold infestation.[79]

Season – The seeds ripen from late summer through late fall and can still be gathered late into the winter. Rhizomes may be harvested year-round but are best from late winter to early spring.

Nutrition – The raw rhizomes of *Nelumbo nucifera* contain 2% protein, 0% fat, 9% carbohydrate, and 1% fiber.[16] They also contain antioxidant phenolic compounds (51 mg GAE/g dw), vitamin C (15 mg/100 g), potassium (2437 mg/100 g), calcium (152 mg/100 g) and magnesium (135 mg/100 g).[16] They are excellent sources of antioxidants and potassium, and moderate to good sources of calcium, magnesium, and vitamin C. These nutrient levels are reduced by cooking, though steaming preserves them better than other cooking methods.[16] The raw seeds of *N. nucifera* contain 16% protein, 2% fat, and 64% carbohydrate, with 342 kcal/100 g.[79] These seeds are also rich in potassium, phosphorus, and magnesium.[79]

Practice – I have not foraged the limited populations of these in Austin.

Nelumbo lutea. Base photo: plants bordering a small pond. Overlay, clockwise from top left: flowers; fruits with mature seeds; range map (GBIF). Photos: GBIF.

TYPHACEAE – Cattail Family

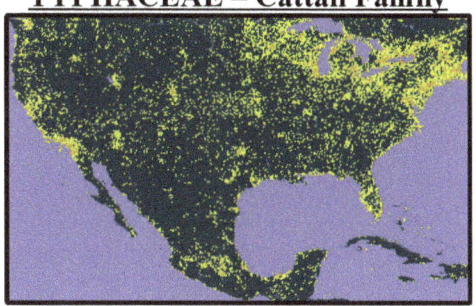 

Typha distribution. Source: GBIF.

Typha spp.
Cattail

Loc.: all TX; not uncommon in Travis Co. (2 spp.); 3 spp. in TX (1 introduced).
Form: grass-like, up to 12 ft. tall; perennial; wet / water habitat.

History

Flowers – When young and green, the inflorescence spike (the cattail) was eaten fresh by the Kawaiisu and Northern Paiute. The Kawaiisu also ate the spike raw when it was brown. Species eaten include *T. domingensis* and *T. latifolia*.

The pollen of *Typha* species was eaten by the Cahuilla, Kawaiisu, and Northern Paiute. To gather it, the inflorescence spikes were shaken or tapped into a basket (Northern Paiute). The pollen was mixed with water to form into cakes, which were baked in coals using layers of cattail leaves to contain them (Cahuilla, Northern Paiute). These cakes could be stored. The pollen was also eaten as mush (Cahuilla).

Seeds – The mature, brown inflorescence spikes of *T. domingensis* and *T. latifolia* were warmed in the sun, and the fluff was rubbed off the spikes (Gosiute, Northern Paiute, Western Natives). The fluff was placed on the ground or on a flat rock in a layer about two inches deep. It was set on fire and the mixture was stirred to burn off all the fluff. The tiny seeds remained on the ground or rock, from where they were swept up. They were then winnowed, ground into meal, and boiled into mush. The dry meal was also eaten with a little water. The fluff may also have been burned off while still on the spike (Gosiute, Western Natives).

Shoots – The emerging shoots of *Typha* species were eaten raw by the Cree, Kawaiisu, Northern Paiute, and Western Natives. A long piece of lateral root (rhizome) was dug from the mud, and the new shoots (culms) emerging from it were broken off (Northern Paiute). The leaves were peeled from these shoots, and the underlying crisp, white tissue was eaten fresh. Shoots from plants growing in moving, rather than stagnant, water were preferred as they were sweeter and better-tasting. These shoots were eaten from early spring through midsummer.

Roots – *Typha* rhizomes were eaten by the Chiricahua and Mescalero Apache, Cahuilla, Cree, Haudenosaunee, Klamath, Northern Paiute, and Western Natives. Species eaten include *T. domingensis* and *T. latifolia*.

The rhizomes were usually three to four inches below the soil surface. They could be eaten at any time but were richest in starch at the end of the growing season. They were gathered in the spring (Apache), at the end of growing season (Western Natives), "late in the season, when full of stored food" (Klamath), just before blooming, or year-round (Cree, Western Natives).

The "young white" rhizomes were preferred (Chiricahua and Mescalero Apache). The outer bark (peel) was first removed, and the central white core, about half an inch in diameter, was eaten raw, scalded, boiled, baked, or prepared by some other cooking method (Chiricahua and Mescalero Apache, Cree, Northern Paiute, Western Natives). The rhizomes were sometimes bruised, and their juice was extracted for a beverage (Haudenosaunee).

The rhizomes were dried for storage, sometimes on a smoking rack (Cahuilla, Cree). They were also split into strips, roasted over a fire, then allowed to air-dry in baskets (Northern Paiute). The dried roots were ground into meal that was cooked with hot stones in baskets (Cahuilla, Northern Paiute). The meal was also moistened to form cakes that were roasted in coals (Northern Paiute).

Notes

Character – Cattails generally grow in perennial water sources, but they are not limited to rivers, lakes, and creeks. They also occur in wetlands and low-lying areas on the landscape that are typically moist or muddy. They can tolerate some seasonal drying.

The species are difficult to distinguish, requiring technical analysis of traits under magnification. For the purpose of foraging, *T. latifolia* and *T. domingensis* are essentially identical, with the Northern Paiute historically using both in the same way for all edible parts.

Typha is most likely to be confused with *Schoenoplectus*, which has leaves that are rounded in cross-section, whereas *Typha* leaves are flattened. The flowering and fruiting heads are also easy to distinguish: only *Typha* has a spike-like (spadix) form, while those of *Schoenoplectus* are branched.

When mature, cattail heads (fruiting spadices) release seeds, each attached to a tuft of fluff that carries it on the wind. Maturity can be assessed by running a hand along the cattail head or bending it. If the fluff extrudes, usually with a light force, the seeds are mature. If not, the cattail will remain intact in a compact form.

Season – The pollen can be gathered from spring through summer. The emerging shoots are also found during this period, but are most abundant in spring. The seeds mature in late summer through fall, often remaining on the stems until early spring. The rhizomes can be harvested year-round, but they are probably best in the fall.

Nutrition – The pollen of *T. latifolia* gathered by the Northern Paiute contained about 1.04 kcal / gram, 5% protein, and 18% carbohydrates (Fowler 1990). Cattail pollen gathering yielded an exceptionally high rate of return for calories consumed from the pollen per calorie expended to gather and process it (Fowler 1990).

Practice – I gather the pollen by tapping the flowering heads into a container. Porous or textured containers will let the pollen escape or stick, so a smooth-surfaced container is ideal. The entire inflorescence can be placed into a plastic or glass bottle and tapped or shaken to release the pollen. It tastes much like regular wheat flour.

I gather the seeds in a similar manner, placing the entire fruiting head into a canvas bag then stroking or bending it to cause the fluff and seeds to be released. A big pile of seed fluff can be set afire in a large metal pot to burn away the fluff and leave the toasted seeds behind.

Additional notes:

The leaves of *Typha* species, which are long, thick, wide, sturdy, and waterproof, were used for mats and basketry by the Caddo, Cahuilla, Flathead, Hocąk, Isleta, Kawaiisu, Mexican Kickapoo, Klamath, Menominee, Meskwaki, Ojibwe, Osage, Gila River Pima, Potawatomi, Tolowa, Yurok, Natives in the Pacific Northwest, and likely others. Cattail leaf mats were important material goods, being used for thatching houses, covering walls, sitting, and sleeping. The Tolowa and Yurok even wove cattail leaves into raincoats.

The Meskwaki and Potawatomi made cattail mats by sewing the leaves together in parallel and binding the edges. Cordage (from nettle or basswood fibers) was used to sew them, being threaded by a calf rib bone needle. Such needles were long, flattened, narrow, and slightly curved. An "invisible stitch" was used, and the mats were sewn with the leaves several layers thick. The edges were whipped tightly with fibers to prevent the from mats unraveling. Such mats were overlapped like shingles for thatching and were impervious to rain and wind.

The fluffy down of *Typha* seeds was used as padding and insulation in pillows, blankets, and bedding by the Caddo, Dakota, Hocąk, Klamath, Meskwaki, Ojibwe, Osage, Omaha, Pawnee, Gila River Pima, Ponca, Forest Potawatomi, Winnebago, Plains tribes and Western Natives. The fluff was specifically used for quilting insulation to keep babies warm in the winter. Pads of the compressed fluff were used by Plains tribes as diapers. The Ojibwe even used the fluff in battle, throwing it into an enemy's face to blind them ("pocket fluff!").

The bright, sun-yellow pollen of cattail flowers was used for ceremonial markings and face paint by the Apache, Diné, and Seri.

Typha domingensis Pers.
Southern cattail

= Typha angustata, T. angustifolia var. dominguensis, T. bracteata, T. truxillensis
Tule
Kawaiisu: *toʔivi*, Northern Paiute: *tahúnadzi*, Gila River Pima: *uḍvak*, Seri: *pat*
Loc.: all TX; not uncommon in Travis Co.; wet / water habitat.
Form: grass-like, up to 12 ft. tall; perennial.
Flowers: Mar-Aug (white, yellow, brown).

Typha latifolia L.
Broadleaf cattail

= Massula latifolia
Tule, *roseau*
Apache: *tel*, Cahuilla: *ku'ut*, Comanche: *pisbu:ni*, Cree: *otawăsk*, Dakota: *wihuta-hu*, Diné: *txatitíi'* – "pollen" / *txyeel* – "broad," Gosiute: *to°ĭmp*, Hocąk: *wicihu* – "leaves mats" / *kšohį* – "baby's coat," Kawaiisu: *toʔivi*, Klamath: *pō'-päs*, Lakota: *wihuta hu* – "tent bottom plant" / *hiŋtkaŋ* – "fuzz, scraped off," Osage: *wakeðe* / *mikeðestsedse*, Menominee: *up'akiuoti'pa*, Meskwaki: *pakwe'ûk* / *piwiê'skînûk*, Ojibwe: (the catkin fuzz) *bebamasûn* – "it flies around" / (leaves) *abûkwe'skwe* – "wigwam cover," Omaha-Ponca: *wahab' igaskonthe*, Northern Paiute: *tóibɨ*, Pawnee: *hawahawa* / *kirit-tacharush*, Forest Potawatomi: *aba'kweûck* – "shelter weed" / *biwie'skwinuk* – "fruit for baby's bed," Prairie Potawatomi: *pakwe'ûk* – "shelter weed," Winnebago: *ksho-hiⁿ*
Loc.: all TX, more C and SE; uncommon in Travis Co.; wet / water habitat.
Form: grass-like, up to 10 ft. tall; perennial.
Flowers: Mar-Aug (yellow, brown, green).
Food: Chiricahua and Mescalero Apache, Cahuilla, Cree, Gosiute, Haudenosaunee, Kawaiisu, Klamath, Northern Paiute, Western Natives.

Typha domingensis. Left: plants in early April. Right: flowering plants (ready for pollen harvest) in mid-May. Top: range map (GBIF). Bottom: fruiting plants in early April (seeds retained over winter).

ASPARAGUS FAMILY

ASPARAGACEAE – Asparagus Family

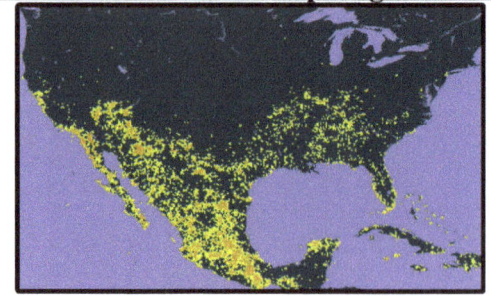

Agave distribution. Source: GBIF.

Agave spp.
Agave

Includes: mescal, century plant, maguey, lechugilla
Apache: *nadah* – "acorns"
Loc.: W, C, & S TX; common ornamental in Travis Co. (5 spp.); 9 spp. in TX (1 introduced).
Form: low shrub, shrub; perennial.

History

Water – To obtain an emergency water source, the tips and margins of *A. cerulata* leaves were trimmed off by the Seri, leaving an approximate rectangular prism. These were roasted over a fire until the outside was charred. The charring was scraped off, and the remaining piece was pounded to extract a sweet juice that was drunk.

Flowers – The flowers of *A. deserti* were eaten by the Cahuilla. They were gathered from April to August. They were parboiled or boiled to eat fresh. The flowers were dried to preserve them, sometimes after being boiled. The dried flowers were boiled to prepare them for eating.

Flowering stalks – The flowering stalks of *Agave* were eaten by the Chiricahua and Mescalero Apache, Cahuilla, 'Iipai-Tiipai, Mexican Kickapoo, Seri, and Tohono O'odham. Species eaten include *A. americana*, *A. deserti*, and *A. parryi*.

They were harvested from April to early summer, when about four to five feet tall and before flowers appeared, being considered too bitter after flowering (Apache, Cahuilla, Tohono O'odham, Seri). They were dug out from the center of the plant with a long stick that had a chisel-edged end (Tohono O'odham). Several hundred pounds of stalks could be gathered in a day in a good area (Cahuilla). They were an important food source (Cahuilla, Tohono O'odham).

They were roasted on coals for about fifteen minutes, after which the outer charred portion was stripped off, and the soft, white, sweet central portion was eaten (Apache). Sometimes, they were eaten raw or boiled (Apache). They were also baked on hot stones to make a tender, asparagus-like food called "*quiote*" (Mexican Kickapoo).

The stalks were peeled, cut into pieces, boiled, and dried for storage (Apache). Roasted stalks were also pounded into cakes, sun-dried, and sealed in pots to preserve them (Cahuilla), or they were cooked immediately and never stored (Tohono O'odham).

Leaves – Not including the base and heart portions (covered below), just the leaves of *Agave* were eaten by the Cahuilla, Havasupai, Hopi, Mayo, and Seri. Species eaten include *A. cerulata* and *A. deserti*.

They were harvested year-round but were considered best from November to May (Cahuilla). Yellowed leaves, or those nearest to the ground, were bitter and not used. The other leaves were cut off, baked, and eaten, or were dried for storage. Roasted leaves were pounded into cakes, sun-dried, and sealed in pots to preserve them.

The tips and margins were trimmed off, leaving an approximate rectangular prism (Seri). These were roasted over a fire until the outside was charred. The charring was scraped off, and the remaining piece was pounded to extract a sweet juice that was allowed to stand for several days to ferment, then was drunk.

The leaves were ground up or broken apart and placed into water, which, in two or three days, yielded a fermented beverage that was drunk (Mayo). Though not mentioned in the Mayo account, it is possible the leaves were first cooked, or perhaps the fermentation process rendered the flesh edible.

Stems & leaves – The basal (inner) portion of the leaves and where all the leaves unite in the center of the plant (basal rosette), often called the "heart" or "crown" of *Agave*, was eaten by the Apache, Cahuilla, Chemehuevi, Hopi, Hualapai, Nebome, Southern Paiute, Seri, Timbasha, Tohono O'odham, Natives in Southern California, Plains tribes, Puebloans in 1540, Indigenous Mexicans, folk in Northern Mexico, Mescaleros in early Mexico, Southwest Natives, Texas Natives 5,000 years ago and in the Archaic, historic Texas Natives, and likely others. Species eaten include *A. americana, A. angustifolia, A. asperrima, A. cerulata, A. chrysoglossa, A. colorata, A. deserti, A. lechuguilla, A. maculata, A. palmeri, A. parryi, A. pelona, A. subsimplex,* and *A. utahensis*.

Agave hearts were gathered from November to December

(Cahuilla), January to February (Cahuilla, Seri), or February to March (Chemehuevi, Southern Paiute). The hearts were gathered when the flowering stalks began to emerge, before they got too tall or began to flower (Cahuilla, Chemehuevi, Southern Paiute, Seri), which happens later at higher latitudes.

The general method of processing was to use a long-handled implement with a sharpened, spatulate head to cut the plants from the ground. A stone blade was used to carefully cut away the leaves near their bases, about a few inches from where they all meet in the center. What was left was the heart, which was baked in earth ovens, usually along with many other hearts. They were cooked for about twelve hours to several days, depending on the number of hearts and the size of the ovens and fires. The cooked hearts were removed and ready to eat or be further processed for storage. A few detailed methods by the peoples in bold text are described below.

The **Apache** processed agave hearts by first removing them out with three-foot-long oak digging sticks with a flattened, spatulate end. The flat end was pounded with a rock into the stem just below the heart to cut the plant from the ground. The leaves were cut off with a stone knife. The hearts were baked in an earth oven, ten to twelve feet in diameter and three to four feet deep. The pit was lined with flat rocks, and a fire was built inside and burned down to coals, which were then removed. Moistened grass was used to line the pit and cover the hearts, which were placed in the pit. A fire of oak and juniper wood was built on top of the oven and kept burning for a day or several days before the hearts were fully cooked. The pulpy center was removed from the charred hearts and was eaten or dried for storage. To dry the hearts, they were pounded into thin sheets on a rock. This juicy pulp was spread out to dry on shallow baskets made from *Yucca* leaves. The juice of the pulp was sprinkled on the drying pulp, forming a glaze. The dried pulp was soaked in water to hydrate it, squeezed to remove excess water, and was eaten plain or mixed with walnut (*Juglans* spp.) meal, juniper (*Juniperus* spp.) fruits, or fragrant sumac (*Rhus aromatica*) fruits. The inner sides of the leaves connected to the heart were chewed and scraped off to eat.

The **Seri** processed agave hearts by first cutting off the spines at the leaf tips, then cutting out the developing inflorescence. Each leaf was then cut off near its base, starting with the inner leaves. An implement with a chisel-like blade at the end that also served as a prybar was pounded in with a rock between the crown and root to sever the two. The leaves were trimmed close with a knife and were discarded as being too bitter. A strip of leaf fiber was left attached to the heart to serve as a carrying strap. The hearts were cooked in an earth oven overnight. The earth oven had a bottom layer of coals upon which the hearts were directly laid, resulting in a charred outer layer. Slices of the heart were eaten fresh with grease or were sun-dried to preserve them, being soaked in water to hydrate them to eat. The remaining leaf bases on the cooked heart were chewed to suck out the juice, and the fibers were discarded, or were pounded on a metate, the fibers were removed, and the pulp was formed into flattened cakes. These cakes were sun-dried to preserve them.

The **Southern Paiute** and **Chemehuevi** used a chisel-shaped wooden implement and a knife to sever the plants from their roots at ground level. The leaves were usually trimmed to within one or two inches from the base (where the leaves meet) with a knife before being loaded onto pack frames. An earth oven was used to cook the hearts for 24 to 48 hours. The cooked hearts were eaten fresh or were cooled, pounded, and formed into large, flat cakes for drying or storing. The cooked heart pulp was also made into a stew with other plant meals or meats. Agave hearts were a primary wild plant food of the Southern Paiute.

The **Cahuilla** gathered the hearts by cutting off the leaves with a shovel-like or spatulate tool made of hard wood, with a sharpened, fire-hardened edge. The plant base was detached by inserting a sharply pointed hardwood pole into it and prying it off. They were cooked in an earth oven for three nights.

The **Tohono O'odham** often obtained *A. americana* hearts by trade but also gathered them. To do so, chisel-shaped digging sticks of hard wood, made by pounding the wood with a stone and hardening the blade in a fire, were used to cut the plant from the ground. The leaves were then cut off with a "mescal hatchet," a semi-circular blade of granite or diorite. The hearts were baked in an earth oven for 12 hours or more. The remaining leaf stubs were pulled off, leaving an inner mass about the size of a cauliflower. It was a sweet and highly esteemed food and was stored in jars.

The Carrizo were eating "maguey" as their main plant food in 1767, likely referring to hearts of *A. americana*. The hearts were baked in earth ovens by Indigenous Mexicans and American Indians, such as the Apache. The hearts were removed and cooked in large earth ovens for three days. The Hualapai preserved the roasted hearts and leaves by pressing them into thin cakes, measuring 30 inches long and 18 inches wide, which they traded to the Hopi for corn. The roasted leaves were bundled and dried by the Apache as a portable food.

A fermented drink was sometimes made from *Agave* (Chiricahua and Mescalero Apache, Tepehuán, other Natives). After the hearts were roasted, the inner part was removed, cut into pieces, pounded into pulp, placed into a hide pouch, and buried for two days or longer. The juice was then squeezed into a container and left to ferment for two or three days, after which it was ready to drink.

The Tohono O'odham and Apache cut the roasted heart into pieces and allowed them to ferment for a week with water in rawhide bags or pottery jugs. The pulp of the hearts was mixed with water and boiled to make a syrup that was dissolved in cold water for a pleasant-tasting beverage.

Mescaleros in early Mexico prepared mezcal in the following manner. When the agave plant was two to four years old, and the stalk was ready to emerge, the plant was cut off close to its roots, stripped of leaves, and the remaining heart was processed. Three to four hundred of these hearts were baked in earth ovens consisting of trenches lined with stones on all sides. The trenches were filled with wood, which was burned into embers, upon which stones were laid to heat up. Agave leaves were layered on the stones, the hearts were placed atop, then were covered with stones, and a fire was set atop it all and kept burning for 8 to 15 days. After the cooking, the hearts were edible and were crushed by foot to extract the sweet liquid that was fermented in hides. Aromatic plants were blended in, especially the bitter root of a plant Berlandier refers to as *raicilla* and a mimosa (Fabaceae), possibly called *mezquitito*. When fermentation was complete and the liquid no longer frothed, it was distilled into liquor.

Mescal heart season in the spring was one of the six time periods the Chiricahua Apache used to divide the year. The cooked hearts had a mild laxative effect. Hearts from *Agave* spp. in the subgenus *Agave* (=*Euagave*) were preferred by the Seri compared to those from the subgenus *Littaea*.

Sap – The sap of *Agave* was eaten by Indigenous people in Tenochtitlan around 1519 and Mexican folk. Species eaten include *A. americana*.

A kind of "honey" (agave syrup) was sold in markets by Indigenous people in Tenochtitlan around 1519. This was likely the concentrated sap of agave, as is still produced in Mexico today. A kind of "sugar" and "wine" from agave was also sold in these markets. The latter was fermented sap, or pulque. Pulque was a common drink in Tenochtitlan when the Spanish first arrived in the early 1500s. It was made from maguey (*A. americana*). It was lightly fermented but was highly regulated if distilled. The distilled pulque was called "aguardiente mezcal." Pulque was produced throughout the historical period of Mexico.

To prepare it, the central shoot was extracted, and the sap that flowed into the pit left was allowed to ferment, then was distilled (folk in Northern Mexico). This was a common commercial beverage, being transported in pigskins or bladders and flavored with fruit juices or citrus peel, and accompanied by salt. The fermented beverage was not produced by American Indians, as most *Agave* species in the United States do not produce enough sap.

Aboveground parts – The Havasupai piled stones on the young sucker plants, forcing them to grow into deformed balls. These balls, as well as the leaves from normal plants, were baked in large earth ovens. This baked food was an important article of trade to the Hopi. The food was soaked in water to yield a drink. The species used was likely *A. parryi*.

Most of aboveground parts of *A. deserti*, including the short, younger leaves around the center, their base, and the flowering stalk, were roasted by the Cahuilla in earth ovens for a day or two, yielding a fibrous, sweet, and nutritious food. Pieces of this were dried and preserved for years. The basal rosette was the most prized part of the plant.

A single *A. deserti* plant might sustain a Cahuilla family for a week, and was a basic food staple. Productive gathering areas of agave were owned by individual family lineages. During dry seasons, the agave crops were less desirable. The leaves were host to the California giant-skipper [*Agathymus stephensi* (Skinner, 1912)] caterpillars, which were roasted on the leaves and eaten by the Cahuilla. Basal rosette collection destroyed individual plants, but suckers sent out by plants to grow young plants around it took their place. Its abundance and the ability to dry and preserve the cooked agave parts made it a highly valuable resource.

Notes

WARNING – **Agave is poisonous** if not properly prepared. The juices of the plant cause a painful burning sensation on the skin, and no part of the plant should be consumed raw.

Character – Agave is a signature plant of the Southwest flora, but is generally not a native wild component of the Austin flora. However, it is commonly used for landscaping in Austin, and these escape into the wild or the populations persist uncultivated. So, they can be found wild in Austin and start to become common in the wild in western Central Texas and westward. The genus is distinct and easily distinguished from yucca or sotol by the succulent, thick, fleshy leaves of *Agave*.

Agave species mostly flower only once before dying (monocarpic), although some species can flower multiple times (polycarpic). Despite the common name "century plant," derived from the idea that they take 100 years to flower, *A. americana* plants typically flower after only ten to thirty years. These thick flowering stalks tower up to thirty feet high, and the plant uses all its energy to produce this majestic inflorescence. Once they die, they are usually replaced by suckers, small plants that are vegetatively reproduced from the root system that are found under and surrounding the mother plant. If one considers these clonal suckers to be branches of the same plant, which they technically are, *Agave* are actually mostly polycarpic. Smaller *Agave* species generally have faster life cycles, roughly proportional to the size difference from *A. americana*.

Mezcal is the distilled liquor from fermented agave heart pulp, and its name comes from the Nahuatl "*mexcalli*," meaning "cooked agave." Tequila is a special type of mezcal that is only made from a specific cultivar of *A. tequilana*.

Season – The hearts are best from winter to early spring, either before or just after the flowering stalks begin to emerge. The flowering stalks are best from spring to early summer, well before they reach full height but after they have emerged above the height of the plant. They are inedible after flowering. The flowers are available from spring through summer. The sap can only be obtained under certain conditions and from certain species, specifically mature *A. americana* plants, though other contexts may be possible.

Nutrition – Cooked agave heart contain 1% protein, 0% fat, 32% carbohydrate (21% sugar), and 11% fiber, with 135 kcal/100 g.[68] The dry cooked flowers of *A. inaequidens* contain 3% protein, 2% fat, 81% carbohydrate, and 8% fiber.[42]

Practice – I use a shovel to cut off the leaves of Agave, then dig up the basal rosette, prying it up and cutting off the roots with my shovel. The leaves can be used as an excellent source of strong fibers, which I have used to make a bow string. I wash off the dirt and remove the lowermost leaves that are fibrous, yellowed, and bitter. I cook the whole thing in an earth oven overnight, but it can also be slow-roasted in a normal oven. The green portion of the attached leaf bases is fleshy, juicy, soft, and sweetish, but have fibers that need to be spat out. The part where all the leaves meet is white and is similar but even sweeter and does not have the same fibers, so I can eat these whole. This white, innermost portion is the true "heart" of agave and the best part. It can be eaten plain or sliced and dried for storage. The heart and the leaf bases can also be pounded into a pulp in a large mortar and put into a large vessel with some water to make an approximation of pulque, though true pulque is from just the sap. This pulp can be strained and drunk, or it can be fermented into an alcoholic beverage. This beverage can be distilled to make tequila or mezcal.

I have also eaten whole leaves and the central leaf stalk. For leaves or the central stalk, I carefully cut them off the plant, wrap them in aluminum foil, then cook them in the oven for about 8-12 hours at about 200° F. Once they are cooked, they are soft, juicy, and sweet. The basal part has lots of soft flesh, but it gets too fibrous toward the leaf tips to be good for eating.

When initially harvesting the leaves, I cut off the prickles along the leaf edges by drawing a knife along the edge. The plant's juices are caustic and cause a severe burning sensation on the skin. I have counteracted this by coating the affected area with mud, providing significant relief.

I have tried to obtain the sap via the historical methods described, but the plant I used was only about five years old, so it may not have been mature enough for the sap to flow.

Agave americana L.
American century plant
= Agave complicata, A. felina, A. gracilispina
American agave, maguey
Tohono O'odham: *a'o't*
Loc.: S, W, & C TX; common ornamental in Travis Co.
Form: succulent, up to 6 ft. tall; perennial.
Flowers: June-July (yellow).
Food: Apache, Carrizo, Hopi, Hualapai, Mexican Kickapoo, Tohono O'odham, Indigenous Mexicans, other Natives, Mexican folk.

Agave lechuguilla Torr.
Lechuguilla
= Agave lophantha var. poselgeri, A. multilineata, A. polselgeri
Tampico fiber
Loc.: W & SW, & C TX.
Form: succulent, up to 2 ft. tall; perennial.
Flowers: May-July (white, pink, yellow).
Food: Nebome, Texas Natives in the Archaic, Mexican folk.

Other *Agave* spp. in Travis Co.:
A. asperrima, *A. lophantha*, and *A. salmiana*
All are mostly ornamental but can be found escaped in the wild.

Agave americana. Base photo: plant in late November. Overlay, clockwise from top left: range map (GBIF); cooked heart; inflorescence in late April; emerging flowering stalk in mid-March.

Agave lechuguilla (in Brewster Co.). Base photo: fruiting plant in early March. Overlay: plants in March and range map (GBIF).

Camassia scilloides (Raf.) Cory
Large camas

= Camassia esculenta (Ker Gawl.) B.L.Rob., Quamasia hyacinthina
Eastern camas, Atlantic camas, common camas, camass, wild hyacinth
Comanche: *siko: / siiko*
Loc.: C & NE, & E TX; not uncommon in Travis Co.
Form: grass-like herb; perennial.

History

Roots & stems – The bulb was peeled and eaten raw by the Comanche. Various archaeological middens (ancient trash pile sites) in central (12 sites) and north (1 site) Texas from 9,000 to 1,000 years ago contained *Camassia scilloides* bulb fragments.[2, 65]

The bulbs of other *Camassia* species were eaten by the Blackfeet, Calpella, Concow, Gosiute, Karuk, Little Lake, Nez Perce, Northern Maidu, Nomlaki, Pomo, Shasta, Tolowa, Wailaki, Yokia, Yuki, Natives of Cape Flattery and Pitt River, and other Natives. All species of *Camassia*, with the exception of *C. cusickii*, were eaten by Natives in the plants' ranges. Species eaten include *C. leichtlinii*, *C. quamash*, and *C. scilloides*.

Among Round Valley Natives, *Camassia* bulbs were dug up with a pointed digging stick made of hard wood. The bulbs were sometimes boiled, but were usually roasted in earth ovens, with large quantities gathered and cooked by several families at once. The raw flesh of the bulb, which was crisp, white, very mucilaginous, and almost tasteless, became quite sweet when cooked. To roast them, they were placed in an earth oven and a fire was built over the hole and kept burning all night and all the next day, by which time they were fully cooked.

The Blackfeet, Gosiute, Shasta, and other Natives also cooked them in an earth oven, for up to two days and nights, and stored them for winter use. After baking in an earth oven, the bulbs were mashed and pressed together into large cakes that were sun-dried for storage, lasting a year or more. The bulbs were also boiled to make a kind of molasses, which was highly prized.

Extreme care was taken to prevent confusing this bulb with the fatally toxic death camas (*Toxicoscordion* spp.), which grows in the same habitat and appears almost identical, but has cream or white color flowers.

Notes – It is possible the species eaten by the Comanche was *Camassia angusta* instead, given that both occur in the study area. Carlson and Jones (1939) did not conclusively identify it, naming it Camassia esculenta (without authorship), which, depending on the authorship, can be a synonym of *C. scilloides* or *C. quamash*, the latter of which does not occur in the study area.

Notes

Character – This species appears similar to the deadly poisonous Nuttall's death camas (*Toxicoscordion nuttallii*), which occurs in the same area. The flowers of Nuttall's death camas are white or cream-colored, whereas those of *C. scilloides* are a lavender or purple color.

It is very difficult to identify *C. scilloides* without the flowers, as it otherwise looks almost like grass, or, at least, like many herbaceous monocot species.

Season – It flowers from late March through early April. The bulbs of other *Camassia* species were gathered from summer through fall.

Practice – I consider this species to be too uncommon to harvest, as gathering the bulbs kills the plants. It is of high ethnobotanical interest, given the importance of the genus as a historical Indigenous food source.

Camassia scilloides. Base photo: flowering patch in early April. Overlay, clockwise from top left: leaves; flowers; young fruits; range map (GBIF).

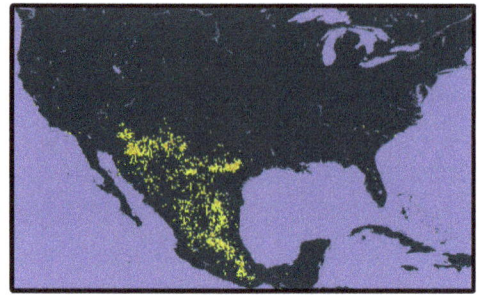

Dasylirion distribution. Source: GBIF.

***Dasylirion* spp.**
Sotol

Loc: W, SW, & C TX; common in Travis Co. (1 sp.); 3 spp. in TX.
Form: low shrub, shrub; perennial.

Food

Fruits & seeds – *Dasylirion* seeds were found to be common in 5,000-year-old coprolites from Natives of Southwest Texas. *D. texanum* seeds were found in Archaic Period earth ovens in Southwest Texas. No historical Indigenous use of the fruits or seeds was found in the references covered, but their use as food in prehistory is apparent.

Flowering stalks – The young flowering stalks of *D. wheeleri* were sometimes eaten by the Chiricahua and Mescalero Apache and the Tohono O'odham. They were dug out from the center of the plant with a long stick with a chisel-edged end. They were roasted on coals for about fifteen minutes, after which the outer charred portion was stripped off, and the soft, white, sweet central portion was eaten. They were sometimes eaten raw or boiled. They were cooked immediately and never stored.

Leaves & stems – The basal rosette of *Dasylirion* species is anatomically equivalent to an agave heart and was eaten by the Chiricahua, Lipan, and Mescalero Apache, Mexican Kickapoo, Tohono O'odham, Texas Natives in the Archaic, other Natives, and Mexican folk. Species eaten include *D. leiophyllum*, *D. texanum*, and *D. wheeleri*.

Dasylirion hearts were a choice food of the Lipan Apache. Large quantities were gathered and cooked for several days in an earth oven measuring three to four feet deep and several feet in diameter. When cooked, they were dried so that the "flakes or thin layers would separate easily." They were then ground into a meal with stone or wood mortars using wood pestles. The meal was mixed with water, formed into small cakes, and cooked in the ashes and embers of a fire.

D. texanum hearts were eaten by American Indians and Mexican folk, either raw or cooked. The hearts were fermented and distilled by

Mexican folk to make "sotol mezcal," which was commonly drunk on the historical frontier. Remains of the stems, leaf bases, and fibrous quids were found in archaeobotanical analyses of Archaic Period earth ovens in Southwest Texas.

D. wheeleri hearts were eaten by the Chiricahua and Mescalero Apache and Tohono O'odham. They were gathered in the spring, when the flowers were beginning to emerge (or perhaps just the flowering stalk). They were dug out with three-foot-long oak digging sticks with a flattened end. The flat end was pounded with a rock into the stem just below the heart to remove it. The leaves were cut off with a stone knife.

The hearts were baked by the Apache in an earth oven that was ten to twelve feet in diameter and three to four feet deep. The pit was lined with flat rocks. A fire was built inside and burned down to coals. The coals were removed and moistened grass was used to line the pit and cover the hearts that were put in. Earth and rocks were used to cover the pit entirely, and a fire of oak and juniper wood was built on top, kept burning for one or several days until the hearts were fully cooked. The pulpy center was removed from the charred hearts, pounded into thin sheets on a rock and spread out to dry. The dried pulp cakes were eaten straight. This preparation method of the Chiricahua and Mescalero Apache was almost identical to their method of cooking agave hearts. However, sotol hearts were considered less desirable since they were more hard and woody, with only the youngest and most tender portions being eaten.

The hearts were sometimes baked by the Apache for only one night, then were peeled, crushed, mixed with a small amount of water in a rawhide container, and allowed to ferment underground for about a day, whereupon it was ready to be drunk. The beverage was sometimes placed in pitch-covered water jars or in wooden jugs cut from trees, being allowed to stand for as long as a month before use. *D. leiophyllum* hearts were also cooked and fermented by the Mexican Kickapoo.

Notes

Character – Sotol plants look very similar to yucca plants, but can be distinguished by the saw-tooth edge of sotol leaves. Only *D. texanum* is found in Austin. It is a common landscaping plant, but also occurs in the wild, where it is uncommon.

The fruits of *Dasylirion* are small, dry capsules with one seed. Given the fibrous nature of the rest of the fruit, the seeds were likely the nutritional target of Texas Natives in the Archaic. No historical or modern source covered in this research mentions their use for food.

Season – The historical gathering seasons of the flowering stalks and hearts were similar to those of agave. The hearts were gathered when the flowering stalks were just beginning to emerge (late winter/early spring), and the flowering stalks were eaten before they reached full size and flowered (spring).

Nutrition – The cooked hearts of *D. cedrosanum* from Mexico contain 0% protein, 1% fat, 6% sugar, and 10% fiber.[19] The seeds of *D. cedrosanum* contain 28% protein, 18% fat, 36% carbohydrate, 16% fiber, iron (5.5 mg/100 g) and zinc (79 μg/100 g).[46] The seeds are a good source of macronutrients and iron.

Practice – I consider the hearts of *D. texanum* to be too uncommon in Austin to harvest, as gathering the hearts kills the plants. It is of high ethnobotanical interest given the importance of the genus as a historical Indigenous food source and the overall commonality of the genus in Texas and the Southwest. The seeds, however, may be a viable foraging target.

Dasylirion texanum Scheele
Texas sotol

Loc.: SW, W, & C TX; common in Travis Co.
Form: low shrub, shrub; perennial.
Flowers: June-July (yellow).
Food: American Indians (hearts), Mexican folk (hearts), Texas Natives in the Archaic (hearts & seeds).

Additional notes:

The dried stems of the flowering stalks of sotol are the best material I have used for creating fire with the friction fire-drill method. It was the preferred material for friction fire-drill spindles and hearth boards historically used by the Chiricahua, Lipan, and Mescalero Apache. Although, both pieces were not necessarily always sotol; oak or yucca flowering stalks were also used. Shredded juniper bark was the preferred tinder.

Sotol (hearth & spindle) fire bow-drill in action.

Dasylirion texanum. Base photo: flowering plant in late July. Overlay, clockwise from top left: range map (GBIF); close-up of leaves (note sawtooth edges); close-up of fruits; fruiting stalk in late July.

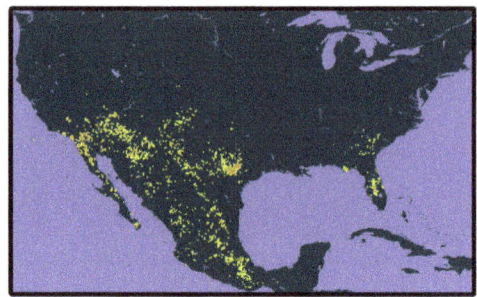
Nolina distribution. Source: GBIF.

Nolina spp.
Beargrass

Loc.: W, SW, C, & sparse SE & NW TX; common in Travis Co.; 6 spp. in TX.
Form: low shrub, shrub; perennial.

History

Fruits & seeds – *N. microcarpa* fruits were eaten fresh or preserved, and the seeds were ground into a meal and eaten (Isleta).

Flowering stalks – *Nolina* flowering stalks were eaten by the Chiricahua and Mescalero Apache and Cahuilla. Species eaten include *N. bigelovii* and *N. microcarpa*.

The Apache sometimes ate *N. microcarpa* flowering stalks. They were roasted on coals for about fifteen minutes, after which the outer charred portion was stripped off, and the soft, white, and sweet central portion was eaten. They were sometimes eaten raw or boiled. The stalks were peeled, cut into pieces, boiled, and dried for storage.

The Cahuilla gathered *N. bigelovii* flowering stalks from May to June and cooked them in an earth oven.

Leaves – The leaves of *N. microcarpa* were the preferred material for lining earth ovens since they do not burn easily (Apache).

Notes

Character – *Nolina* plants look like a large clump of grass with very long blades. The leaves of *N. lindheimeriana* are flattened in cross-section, whereas those of *N. texana* are rounded. The edges of *N. lindheimeriana* leaves are lined with tiny serrations.

Beargrass flowering stalks are anatomically similar to those of sotol or yucca. Those of *N. texana* are much smaller, usually not reaching above the clump of leaves, and can be oriented to the side, whereas *N. lindheimeriana* flowering stalks are several times taller than the leaf clumps and orient upwards, above the plant.

Season – *N. lindheimeriana* flowers from April through June and *N. texana* flowers from March through July.

Practice – I have not tried eating either of the local species of Nolina. The flowering stalks are smaller than those from the species that were historically eaten, especially those of *N. texana*.

Nolina lindheimeriana (Scheele) S. Watson
Devil's shoestring

Ribbon grass
Loc.: C TX; common in Travis Co.
Form: low shrub, shrub; perennial.
Food: no records.

Nolina texana S. Watson
Texas sacahuista

Texas beargrass, basket grass, bunchgrass
Loc.: SW, W, C, & sparse SE TX; common in Travis Co.
Form: low shrub, shrub; perennial.
Food: no records.

Nolina lindheimeriana. Left: fruiting stalk in early August. Right: leaves of same plant and range map (GBIF). Note the similarity to *Yucca*, but the leaves are longer and thinner.

Nolina texana. Base photo: fruiting plant in late April. Overlay, clockwise from top left: flowering stalk in late March; close-up of flowers; range map (GBIF); close-up of fruits and leaves.

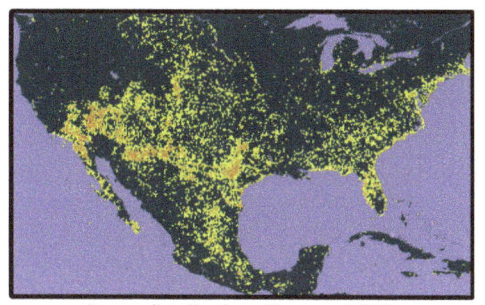

Yucca distribution. Source: GBIF.

Yucca spp.
Yucca

= Sarcoyucca
Diné: *tshá'aszi'*, Northern Ute: *wi̠si*, Zuñi: *ho'kĭapa* – "long leaf wide"
Loc.: all TX; very common in Travis Co. (6 spp.); 18 spp. in TX.
Form: low shrub, shrub, tree; perennial.

History

Flowers – The flowers of some *Yucca* species were eaten by the Chiricahua and Mescalero Apache, Cahuilla, Kiowa, Southern Paiute, Timbasha, Mojave Desert Natives, American Indians, and Mexican folk. Species eaten include *Y. angustissima*, *Y. baccata*, *Y. brevifolia*, *Y. elata*, and *Y. glauca*.

Young flowers were parboiled and mature flowers were boiled up to three times, changing the water in between, with salt added, and were dried to preserve them (Cahuilla). Flowers were also simply boiled (Apache, Cahuilla). The flower buds of *Y. brevifolia* were roasted on hot coals (Timbasha, Southern Paiute) and the flower buds of *Y. baccata* were also eaten (American Indians, Mexican folk).

Fruits – The fruits of some *Yucca* species were eaten by the Chiricahua, Mescalero, Western, and other Apache, Cahuilla, Diné, Hopi, Isleta, Kawaiisu, Kiowa, Southern Paiute, Pima, Seri, Taos, Tewa, Timbasha, Tohono O'odham, Zuñi, Southwest Natives, Texas Natives in the Archaic, and Mexican folk. Species eaten include *Y. angustissima*, *Y. baccata*, *Y. brevifolia*, *Y. elata*, *Y. faxoniana*, *Y. glauca*, *Y. schidigera*, *Y. thompsoniana*, and *Y. treculeana*. Of these, *Y. baccata* (datil) was, by far, the most commonly eaten, followed by *Y. glauca* (Diné, Isleta, Kiowa, Tewa, Zuñi).

Y. baccata fruits had a sweet pulp that was eaten raw, dried, or cooked and was gathered in large quantities and stored. *Y. faxoniana* fruit pulp was also eaten similarly (Apache, Hopi, Tewa). The whole fruits of *Y. glauca* were boiled and eaten plain, but only the immature ones, as the mature pod seeds were not edible (Zuñi). The preparation methods described for the other species were limited, but it is likely that in general, either the mature pod pulp was eaten or the immature fruits were boiled.

The fruits of *Y. treculeana* were gathered in November and December and eaten by Mexican folk. The fruits of other species were gathered in July and August (Western Apache).

The seeds of *Y. baccata* were stored and ground into meal by the Tohono O'odham to eat.

Flowering stalks – The flowering stalks of some *Yucca* species were eaten by the Chiricahua and Mescalero Apache, Cahuilla, Kawaiisu, Kiowa, Southern Paiute, Timbasha, Mojave Desert Natives, folk in Northern Mexico, and Texas Natives. Species eaten include *Y. angustissima*, *Y. baccata*, *Y. brevifolia*, *Y. glauca*, and *Y. treculeana*.

The flowering stalks were gathered in the spring. They were harvested before flowering (Apache, Cahuilla, Texas Natives), just after flowering (Apache, Kawaiisu), or when young and newly emerging (Southern Paiute, Timbasha, Mojave Desert Natives). They were cooked in earth ovens (Cahuilla, Southern Paiute, Timbasha), baked in coals (Mexican folk), or were sliced, parboiled, and cooked like squash (Cahuilla). They were also roasted on coals for about fifteen minutes, after which the outer charred portion was stripped off, and the soft, white, and sweet central portion was eaten (Apache). They were sometimes eaten raw or boiled (Apache). The stalks were peeled, cut into pieces, boiled, and dried for storage (Apache). Before developing the inflorescence, the tips of the flowering stalks of *Y. treculeana*, called "*chiote*," were eaten by Texas Natives.

Leaves & stems – The tender central part of the leaves of *Y. baccata* was cooked in soups and boiled with meat (Chiricahua and Mescalero Apache). The portion of the stem from the ground to the leaves of *Y. elata*, or the "crowns"/"hearts," similar to those of agave, was peeled and baked overnight in an earth oven (Chiricahua and Mescalero Apache). These hearts were then dried in the sun, broken into pieces, softened in water, and eaten. The hearts of *Hesperoyucca whipplei* (Torr.) Baker (= Yucca whipplei) were cooked in earth ovens and eaten by the Cahuilla, Kawaiisu, Southern Paiute, and Timbasha.

Notes

Character – *Yucca* is distinguished from other genera in the Asparagaceae by its wide, flat leaves, large, bell-shaped, cream-colored flowers, and relatively large fruit pods. *Y. rupicola* is by far the most common *Yucca* species in Austin. Of the *Yucca* species with fruits that were historically eaten, only *Y. treculiana* (Spanish dagger) is found in Austin, where it is a common ornamental that is also sometimes found in the wild.

Season – *Y. rupicola* flowers from April to June. Its flowering stalks emerge in April.

Nutrition – The dried fruit pulp of *Y. baccata* contain 3% protein, 1% fat, 86% carbohydrate, 7% fiber, and 417 kcal/100 g.[75] It also has potassium (826 mg/100 g), calcium (370 mg/100 g), iron (52 mg/100 g), zinc (25 mg/100 g), copper (2 mg/100 g), and vitamin A (36 μg RAE/100 g).[75] These nutrients make them very carbohydrate-rich, a good source of fiber, potassium, and calcium, and an excellent source of iron, zinc, and copper. The dried fruit pulp of *Y. angustissima* contain 6% protein, 9% fat, 62% carbohydrate, 7% fiber, and 401 kcal/100 g.[75] It also has potassium (711 mg/100 g), calcium (626 mg/100 g), iron (4 mg/100 g), zinc (2 mg/100 g), copper (1 mg/100 g).[75] So, these are also very energy-dense and good to excellent sources of all of these nutrients.

The nutritional content of dried cooked *Yucca* flowers is likely similar to those from *Agave*, which had 3% protein, 2% fat, 81% carbohydrate, and 8% fiber.[42] The flowering stalks are also likely mostly carbohydrate and fiber, with small amounts of protein and fat.

Practice – I have eaten the flowers of *Y. rupicola* and various other *Yucca* species. Raw, the flowers leave a kind of unpleasant, slightly caustic sensation in the throat. They otherwise taste fine, but it is probably not advisable to eat them raw. They become much better when cooked in any way. Boiling them is the simplest way to cook them, but they can be made very delicious by battering and frying them to make yucca flower fritters.

The flowering stalks are best eaten when they are just emerging above the height of the rest of the plant, or are about two feet tall. They look like a giant asparagus. I cut them off at the base, wrap them in foil, and cook them on low in the oven for several hours. I have not tried eating the fruits of *Y. treculiana* or the hearts of any species. *Y. arkansana* fruit pulp may be edible, as it was formerly classified as a variety of a species that was eaten, *Y. angustissima*.

Additional notes:

Yucca plants were historically a common fiber and soap source among Indigenous peoples. The roots and other parts of all *Yucca* species can be pounded up and lathered in water to use as an effective soap that is useful for sanitary food preparation. I isolate the fibers for cordage by boiling the leaves, chewing them and spitting out the extraneous material, washing the fibers, and scraping the last bits off with my thumbnail.

Yucca arkansana Trel.
Arkansas yucca

= Y. angustissima var. mollis
Loc.: all TX except W, NW, & far N; uncommon in Travis Co.
Food: no records.

Yucca rupicola Scheele
Twisted-leaf yucca

Texas yucca
Loc.: C TX; very common in Travis Co.
Form: herb, low shrub, shrub; perennial.
Food: no records.

Yucca treculeana Carrière
Spanish dagger

= Yucca canaliculata, Y. macrocarpa
Don Quixote's lace, *palmito, pita*
Loc: S TX; common (cultivated) in Travis Co.
Form: herb, shrub, tree; perennial.
Food: Apache (fruits), Texas Natives (flowering stalks), Mexican folk (fruits).

Yucca arkansana. Emerging flowering stalk in early April. Range map: GBIF.

Yucca rupicola. Base photo: plant in late March (note twisted leaves). Overlay, clockwise from top right: flowers in mid-May; emerging flowering stalk in mid-April; range map. Range map: GBIF.

Yucca treculiana. Plant in mid-June. Range map: GBIF.

PALM FAMILY

ARECACEAE – Palm Family

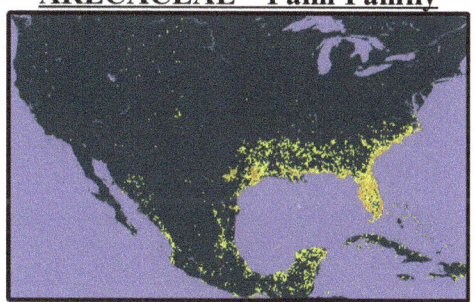

Sabal distribution. Source: GBIF.

Sabal spp.
Palmetto

Loc.: E, S, & C TX; common in Travis Co. (2 spp.); 3 spp. in TX.
Form: shrub, tree; perennial.

History

Fruits – *Sabal mexicana* fruits were eaten by the Mexican Kickapoo. These fruits were historically sold for food in markets of Brownsville and Matamoros, Texas, being known as "*micharos*" there.[72] The fruits were also eaten by members of the 1680s La Salle expedition to Texas. The fruits of an unknown species of *Sabal* were eaten by Texas or Southeast Natives in 1530, as implied by Cabeza de Vaca's account. The fruits of *S. uresana* in Sonora were eaten fresh by the Seri.
Leaves – The young, embryonic fronds of *Sabal* species were cut out of the top of young plants and boiled to eat by Southeast Natives.
Trunks – The inner part of the trunks of Mexican palmetto was dried and pounded into a meal by the Mexican Kickapoo.
Roots – The Houma sliced, baked, and and ate the fresh roots of *S. minor*.

Notes

Character – This genus contains the only native palms in Central Texas. The fruits of Mexican palmetto are an incredibly underrated food. They typically go to waste in Austin and the southern half of Texas, where this plant is commonly used for landscaping and is not uncommon in the wild, especially near creeks. The fruits are borne in large clusters of hundreds of fruits, each composed of about half semi-dry flesh and half a single large seed. They taste sweet, like dates, are soft and slightly chewy, and can be gathered in huge volumes.

The fruits of dwarf palmetto have a thin layer of flesh surrounding a single large seed. They taste very similar to Mexican palmetto, but do not furnish nearly as substantial a food, with the infructescences only bearing dozens of the fruits.

The fronds of both species are similar, distinctly palm-like, and large, but those of *S. mexicana* are significantly larger and are often somewhat folded longitudinally, whereas those of *S. minor* are more flattened and fan-like. The presence of a large trunk characterizes *S. mexicana*, whereas *S. minor* either has no trunk or a small trunk that is usually only a foot or two tall.

Season – The fruits begin ripening in fall and can be found until spring.
Nutrition – The flesh of fresh, ripe *S. mexicana* fruits from Veracruz contain 4% protein, 56% carbohydrate, 25% fiber, and 2615 kcal/kg.[45]
Practice – *S. mexicana* fruits are one of my favored foraging targets as they are easy to gather in great abundance, very delicious, and substantial fresh and raw. I simply gather them by hand, but a ladder helps, as the infructescences can sometimes be difficult to reach on taller individuals. They can be gathered from the ground, but given the urban habitat typical of the ornamental plants from which I often gather, I prefer to harvest them directly from the stems. I keep the fruits in a dry environment, but do not usually bother to directly dry them, as they seem to retain their quality and not rot when not dried. It is, however, probably better to dry them fully for long-term storage. I usually eat them whole and plain, spitting out the seeds. If the flesh of many are freed from the seeds, they would make an excellent substitute for dates in recipes. *S. minor* fruits make an interesting passing snack in the field, but are not worth serious efforts to harvest in abundance.

I have tried digging out *S. minor* roots but they are quite deep. It is possible the Houma were eating the hearts of the underground stem.

Sabal mexicana Mart.
Mexican palmetto

= Sabal exul, S. texana
Rio Grande palmetto, Texas palmetto, Texas palm, Texas sabal, *palma huichira*
Loc: SE, S, & C TX; not uncommon in Travis Co.; common ornamental.
Form: shrub, tree, up to 50 ft. tall; perennial.

Sabal minor (Jacq.) Pers.
Dwarf palmetto

= Sabal adansonii, S. deeringiana, S. glabra, S. louisiana, Corypha minor
Bluestem palmetto, *palamito*, *latanier*
Loc: E, SE, & C TX; common in Travis Co.
Form: shrub; perennial.

Sabal minor. Base photo: fruiting plant in late December. Overlay: close-up of harvested fruits in late September; range map (GBIF).

Sabal mexicana. Base photo: large plant in late September. Overlay, clockwise from top left: range map (GBIF); ripe fruit cut in half; leaf of small plant; infructescence in late February.

GRASS

POACEAE – Grass Family

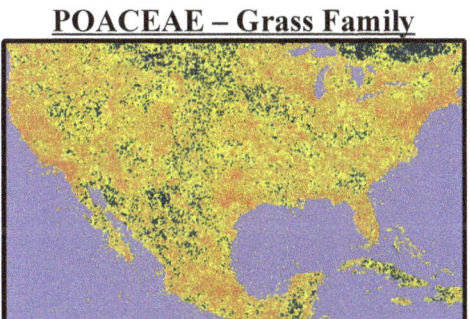

Poaceae distribution. Source: GBIF.

Grass

Flathead: *supuʔlexw*, Hopi: *tu:'saka*, Tewa: *ta*, Northern Ute: *pau-waa=pɨ*
Loc.: all TX; very common in Travis Co.; over 600 spp. in TX.

History

Seeds – All grass seeds are technically edible, but most would require more effort to process than they would be worth given the amount of nutrition they contain. In general, the larger-grained grasses that are the easiest to process were used for food by Indigenous peoples.

Grass grains were an important food source of the Chiricahua, Mescalero, and Western Apache, Cahuilla, Gosiute, Seri, Tohono O'odham, and Natives in Southern California. Grass grains were also eaten by the Apache, Asinai, Assiniboine, Caddo, Calpella, Concow, Diné, Haudenosaunee, Hocąk, Hopi, Isleta, Karuk, Kawaiisu, Kiowa, Klamath, Little Lake, Menominee, Meskwaki, Nomlaki, Ojibwe, Pima, Pomo, Potawatomi, Salteaux, Sauk, Seneca, Sioux, Timbasha, Wailaki, Yokia, Yuki, Yuma, Zuñi, Great Basin Mojave Desert Natives, Great Lakes Natives, Arizona Natives, California Natives, and Texas Natives in the Archaic. Wild species eaten include *Achnatherum hymenoides*, *Agrostis perennans*, *Avena fatua*, *A. sativa*, *Bouteloua barbata*, *Bromus marginatus*, *B. diandrus*, *B. hordeaceus*, *B. tectorum*, *Cinna arundinacea*, *Elymus* sp., *E. canadensis*, *E. elymoides*, *E. glaucus*, *E. multisetus*, *E. repens*, *Eriocoma hymenoides*, *Eriochloa sericea*, *Festuca octoflora*, *Glyceria* spp., *G. fluitans*, *G. striata*, *Hopia obtusa*, *Hordeum jubatum*, *H. murinum*, *H. pusillum*, *H. vulgare*, *Leymus triticoides*, *Melica imperfecta*, *Muhlenbergia* spp., *M. microsperma*, *Panicum* spp., *P. bulbosum*, *P. capillare*, *P. urvilleanum*, *Phalaris caroliniana*, *P. minor*, *Phragmites* spp., *P. australis*, *Poa* spp., *Rostraria cristata*, *Setaria* spp., *S. macrostachya*,

Sorghum halepense, *Sporobolus* spp., *S. airoides*, *S. contractus*, *S. cryptandrus*, *S. flexuosus*, *S. giganteus*, *S. wrightii*, *Zizania aquatica*, and *Z. palustris*.

Cultivated species eaten include *Hordeum vulgare* (barley), *Sorghum bicolor* (sorghum), *Triticum aestivum* (wheat), and *Zea mays* (corn/maize). Maize was the most important cultivated food plant of Indigenous peoples in pre-contact North America.

The Chiricahua Apache made wild grains into a flour for gravy and bread. Grass seeds of many species were eaten by the Cahuilla during periods of scarcity. Small seeds, probably mostly from grasses, were estimated to comprise 14% of the diet of California Natives.

Wild grass grains were generally gathered from early summer to fall by striking the seed heads with a basketry racket into a large-mouthed basket or other receptacle (Apache, Cahuilla, Gosiute, Karuk, Kawaiisu, Kashaya Pomo, Round Valley Natives, Timbasha). The fruiting heads were also cut, bundled, dried, and threshed to release the grains (Apache, Kawaiisu, Pima, Seri, Tohono O'odham). Bunches of grass were sometimes held near a fire, allowing the seeds to fall out to the base of an adjacent flat stone placed obliquely (Diné). In another technique, bundles of fruiting grass were burned to release the grains (Kawaiisu). Another method was to burn a whole patch of grass and sweep up the seeds from the ground (Tohono O'odham). Bristly spikes of barley grains were burned off before winnowing and grinding (Kawaiisu).

Before storage, the grains were winnowed in a tray by rubbing the chaff loose, then tossing them into the air (Kashaya Pomo, Tohono O'odham). The whole grains were also ground into a meal first and then the remaining chaff was winnowed or sifted out (Timbasha). Grains were also lightly pounded in a bedrock mortar to loosen the seeds from the chaff, which was winnowed off on a basketry tray (Kawaiisu).

The hulls of the grains were parched off, usually by shaking them in a shallow basket with live oak coals (Caddo, Cahuilla, Karuk, Pima, Kashaya Pomo, Tohono O'odham, Round Valley Natives). The chaff was sometimes winnowed at this point instead (Karuk). Then the seeds were ground to a fine powder in a mortar or on a metate (Caddo, Cahuilla, Diné, Karuk, Pima, Kashaya Pomo, Round Valley Natives, Yuma). In one method, grains were ground in a mortar, then put into a shallow, obtusely conical basket and swirled to carry the heavier particles to the top of the basket, which were removed and reground (Round Valley Natives).

Various other seeds made into piñole were stored separately, then mixed for different flavors when one was ready to eat, sometimes with nuts and dried fruits added (Kashaya Pomo, Round Valley Natives). The grain meal, sometimes with other seeds, was also cooked into a mush or gruel (Apache, Cahuilla, Isleta, Kashaya Pomo, Karuk, Kawaiisu, Seri, Arizona Natives). Sometimes, grass seeds were cooked into a gruel without being ground up first (Cahuilla). The seed meal was also formed into cakes that were cooked (Diné, Isleta, Kashaya Pomo, Zuñi, Arizona Natives). Dumplings, rolls, griddle-cakes, tortillas, and biscuits were also made (Diné, Kiowa). The flour could also be simply eaten plain and dry (Kawaiisu, Pima) or drunk mixed with water (Round Valley Natives). Wild grain flour was also used for bread, gravy, and to thicken stews (Apache). It was also mixed with cornmeal for various foods (Hopi).

Wild rice (*Zizania aquatica* and *Z. palustris*) was an important food source in the area these plants were found (Assiniboine, Haudenosaunee, Hocąk, Menominee, Meskwaki, Ojibwe, Potawatomi, Salteaux, Sauk, Sioux, other Natives). These are aquatic plants and the grains were gathered by canoe. Canoes were poled in lakes full of wild rice, and the passenger bent the grass stems over the canoe using a curved stick and struck the grains off into the canoes with a small paddle. Forty pounds were gathered at a time, then the grains were placed into a metal washtub tilted over a fire and stirred continuously with a wood paddle until they were parched. The grains were alternatively spread on a lattice rack to dry, with a slow fire kept burning below. Such racks could measure four feet high, eight feet wide, and twenty to fifty feet long, and were covered with reeds and grass. Rack-drying took thirty-six hours and the point was to cause separation of the husk for removal.

After drying, they were hulled by threshing. One threshing method was to make a hole in the ground about one foot deep and one foot wide, line it with skins, fill it with about two dry gallons of rice grains, and step in it with a jumping movement, one foot at a time alternating, to remove the husks. The person doing the trampling was wearing new moccasins. In another threshing method, a blanket, rush mat, or deerskin was spread on the ground, and the dried grains were placed upon it. Three sides of this were covered vertically with the same material, forming a box with an open top and front, from where the grains were beaten with straight sticks in each hand, releasing the hulls. It was then winnowed to remove the chaff. It was often eaten with wild fowl or other game, with maple sugar added, or just with maple sugar to make a pudding.

Note that the only other *Zizania* species in North America is *Z.*

texana, which is endangered and endemic to the headwaters of the San Marcos River in Central Texas. It is possible that this species was used for food by Natives in the area.

Other species that were important food sources include *Achnatherum hymenoides*, *Avena fatua*, *Eriocoma hymenoides*, *Eriochloa sericea*, *Hopia obtusa*, *Rostraria cristata*, *Sporobolus airoides*, *S. contractus*, *S. cryptandrus*, and *S. giganteus*.

Notes

Character – Of the grass species historically eaten, the ones that are occur in Travis County (and Central Texas) and are not rare are *Avena fatua*, *Bromus diandrus*, *Elymus canadensis*, *Festuca octoflora*, *Hordeum murinum*, *Muhlenbergia* spp., *Panicum* spp., *Poa* spp., *Setaria* spp., *Sorghum* spp., and *Sporobolus* spp. Of these, *Avena fatua* likely is the best foraging target, being common, introduced, widely used for food historically, large-grained, and relatively simple to process.

Season – Grass grains ripen at different times depending on the species, but generally are available from early summer through fall. *Avena fatua* begins to ripen in early summer. The ripeness of any species can be evaluated. If the grain heads are dry and readily release the grains when struck, they are ripe.

Nutrition – *Avena fatua* seeds contain 15% protein, 12% fat, and 3% ash (Bean and Saubel 1972).

Practice – I have gathered *Avena fatua* by striking the grain heads with a stick into a container. They can be rubbed between the hands to loosen the chaff, which can be winnowed off. I parch them by shaking them in a basket with coals and grind them in a stone mortar and pestle before eating the meal plain and dry or adding water and cooking it into a mush.

Avena fatua L.
Wild oat

= Avena byzantina, A. sativa var. fatua
Spring wild oat, *avena*
Diné: *łii' pitł'oh naatqa'* – "horse, his grass, corn" / *łii' pitqa'* – "horse, his food," Gosiute: *o'-a-tûmp*, Kawaiisu: *sikʷeʔevi*, Gila River Pima: *koksham* – "it has a coat or shell" / *aatoks muḑadkam* – "it has tassels hanging down"
Loc.: all TX except far S & NE; common in Travis Co.; introduced; 2 *Avena* spp. in TX (*A. sativa*).
Form: grass; annual.
Food: Cahuilla, Calpella, Concow, Karuk, Kawaiisu, Little Lake, Nomlaki, Pomo, Wailaki, Yokia, Yuki, California Natives.

Bromus diandrus Roth
Ripgut brome
= Bromus rigidus (= *B. diandrus* ssp. *rigidus*)
Karuk: *aktipannara*
Loc.: C & E TX; uncommon in Travis Co.
Form: grass; annual, perennial.
Food: Karuk.

Elymus canadensis L.
Canada wildrye
Cahuilla: *pá-han-kis*, Gosiute: *o'-ro-rop*
Loc.: all TX; not uncommon in Travis Co.; 15 *Elymus* spp. in TX (1 introduced).
Form: grass; perennial.
Food: Gosiute.

Eriochloa sericea (Scheele) Munro ex Vasey
Texas cupgrass
= Helopus junceus, Panicum sericatum, Paspalum racemosum, P. sericeum
Loc.: SW, S, C, N & E TX; uncommon in Travis Co.
Form: grass; perennial.
Food: Texas Natives in the Archaic.

Festuca octoflora Walter
Sixweeks fescue
= Diarrhena setacea, Festuca setacea, Vulpia octoflora
Loc.: all TX except far S; uncommon in Travis Co.; 4 *Vulpia* spp. in TX.
Form: grass; annual.
Food: Gosiute.

Hordeum murinum L.
Mouse barley
= H. glaucum, H. leporinum (= *H. murinum* ssp. *leporinum*), H. stebbinsii, Critesion murinum
Farmer's foxtail, foxtail barley, wall barley, hare barley
Cahuilla: *pa'ish heqwas* – "field mouse's tail," Gila River Pima: *ba'imuḑkam* ~ "tassels will stick in the throat" / *koson bahi* – "packrat's tail"
Loc.: W, C, & sparse N & E TX; uncommon in Travis Co.; introduced.
Form: grass; annual.
Food: Cahuilla, Calpella, Concow, Little Lake, Nomlaki, Pomo, Wailaki, Yokia, Yuki.

Muhlenbergia spp.
Muhly
Diné: *pé'ejoo'* – "broom/brush/comb"
Loc.: all TX except far S; very common in Travis Co.; 52 spp. in TX (1 introduced).
Form: grass; annual, perennial.
Food: Apache.

Panicum spp.
Panicgrass

Cherokee: *kanesgá*
Loc.: all TX; common in Travis Co.; 22 spp. in TX (5 introduced).
Form: grass; annual, perennial.
Food: Apache, Cahuilla, Diné, Hopi, Arizona Natives.

Poa spp.
Bluegrass

Loc.: all TX; common in Travis Co.; 16 spp. in TX (5 introduced).
Form: grass; annual, perennial.
Food: Great Basin Mojave Desert Natives.

Setaria spp.
Bristlegrass

Loc.: all TX; common in Travis Co.; 16 *Setaria* spp. in TX (6 introduced).
Form: grass; annual, perennial.
Food: Caddo, Seri.

Sorghum halepense (L.) Pers.
Johnsongrass

= Holcus halepensis
Kiowa: *soñ-toñ-pa / soñ-ka*, Gila River Pima: *kaañu chu'igam* – "cane sorghum, it looks like" / *vaapk chu'igam* – "carrizo / giant reed, it looks like"
Loc.: all TX; very common in Travis Co.
Form: grass; perennial.
Food: Pima.

Sporobolus spp.
Dropseed

Loc.: all TX; not uncommon in Travis Co.; 23 spp. in TX (1 introduced).
Form: grass; annual, perennial.
Food: Apache, Diné, Hopi, Kawaiisu, Kiowa, Tohono O'odham, Great Basin Mojave Desert Natives, Texas Natives in the Archaic.

Avena fatua. Flowering plants and close-up of flowers in late March. Range map: GBIF.

Bromus diandrus flowering plants in late March. Range map: GBIF.

Elymus virginicus fruiting plants in mid-April. Range map: GBIF.

Muhlenbergia lindheimeri fruiting plants in late October. Range map (*Muhlenbergia* spp.): GBIF.

Panicum capillare (witchgrass) fruiting plant in mid-September. Range map (*Panicum* spp.): GBIF.

Sorghum halapense. Base photo: plants in early March. Overlay: flowers in mid-September. Range map: GBIF.

TOXIC

Cicuta maculata L.
Spotted water hemlock

= Cicuta curtissii, C. mexicana
Cherokee: *tiliyustí* – "chestnut-like" / *kanàsolá*, Cree: *maciskatask*, Klamath: *skä'-wänks*, Ojibwe: *apagwasî'gons*, Osage: *žoŋxaštoŋga* (*žoŋ* – "wood", *toŋga* – "large")
Loc.: all TX except far S & far W; not uncommon in Travis Co.; only *Cicuta* sp. in TX.
Form: herb; biennial, perennial.
Family: Apiaceae (Parsley Family).
WARNING – DEADLY POISONOUS (all parts).
Notes – The roots were mashed, mixed with venom from a rattlesnake's venom sacs, or the decomposed liver of a deer or other animal that had been buried for a few days, and used as a poison for Klamath war arrows.

Conium maculatum L.
Poison hemlock

Loc.: C & sparse E, N, & W TX; common in Travis Co.; only *Conium* sp. in the US; introduced.
Form: herb; biennial.
Family: Apiaceae (Parsley Family).
WARNING – DEADLY POISONOUS (all parts).

Left: *Cicuta maculata* (photo from GBIF) Note similarity to *Conium maculatum*.
Right (4 images): *Conium maculatum* stems (note purple splotches); parsley-like leaves; flowers.
Range maps: GBIF.

Colocasia esculenta (L.) Schott
Taro

= Arum esculentum, Colocasia antiquorum
Coco yam, elephant ears, *talo, taloes, colocasi, gölevez, dasheen, inhame, malanga, gabi, madumbi, boina, magimbi, jimbi, eddo, arvi, pheuak, toran, qulqas, kocu, yù, satoimo*
Hawaiian: *taro / kalo*
Loc.: E, C, & sparse N, S, & W TX; common in Travis Co.; introduced; only *Colocasia* sp. in US.
Form: herb; perennial; aquatic / wet habitat.
Family: Araceae (Arum Family).
Notes – considered invasive in the Southeast US.

WARNING – DEADLY POISONOUS. Consumption of any part of wild taro may be fatal. Like other species in the Araceae, the tissues contain calcium oxalate raphides: microscopic, needle-like structures that cause mechanical damage to cells of the digestive tract. These compounds are tasteless. Effects may not be immediately apparent but can produce intense burning in the mouth and digestive tract. The microscopic needles can cause massive tissue damage, being perceived as a sharp, painful, raw burning sensation in the mouth and anywhere the tissue contacts. It can take as long as ten minutes after contact for this sensation to be perceived.

Wild taro contains far higher concentrations of raphides than cultivated varieties. In my experience, vigorously boiling wild taro corms for over 25 hours is necessary before the raphides became less detectable by way of holding the cooked root in my mouth for at least 10 minutes. Even then, they still had dangerous levels of raphides. I have also experimented with wild corms by thinly slicing them, cooking these slices until dry, powdering them, then further dry-roasting the powder. This powder had almost undetectable raphides, but I would still consider the plant to be extremely dangerous to eat under any circumstances. Merely cutting up the root caused my hands and lower arms to painfully burn for hours. Inhaling the powdered root or getting juice in the eyes would also be extremely dangerous.

Notes – In cuisines of cultures around the world, the leaves were commonly eaten cooked, often as edible wrappings, and the stems and leaf buds are also used.

The cooked corms feature prominently in many cuisines globally. Taro is one of the most ancient cultivated foods and remains a staple crop of African, Oceanic, East Asian, Southeast Asian, and South Asian cultures. Corms were typically roasted, baked, or boiled. The plants were not eaten raw, only cooked. Various cooking methods can remove raphides from *cultivated taro* varieties, but **these methods will not make the *wild taro* varieties safe for consumption**.

Datura wrightii Regel
Sacred thorn-apple

= Datura metel var. quinquecuspida
Sacred datura, angel's trumpet, desert thorn-apple, downy thorn-apple, Jimson weed, Jamestown weed, Indian apple, prickly datura, *toloache*
Apache: *itanasbase* – "round leaf," Cahuilla: *kí-ki-sow-il*, Diné: *ntíɢíliitshoh* – "sunflower, large," Hopi: *shemóna / tcimo'na*, Kawaiisu: *moopi*, Zuñi: *a'neglakya*
Loc.: all TX except far E; common in Travis Co.
Form: herb, low shrub; annual, perennial.
Family: Solanaceae (Nightshade Family).
WARNING – DEADLY POISONOUS (all parts).

Left: *Colocasia esculenta*. Note the "elephant ear" shape of large leaves.
Right: *Datura wrightii*. Note thorny fruit.
Range maps: GBIF.

Dermatophyllum secundiflorum (Ortega) Gandhi & Reveal
Texas mountain laurel

= Sophora secundiflora, Broussonetia secundifora, Calia erythrosperma, C. secundiflora
Frijolillo, colorín
Apache: *yułtudi* – "red bead," Comanche: *aincapu*, Kiowa: *k'awn-k'odl / ʔkawn-ʔkodl / kʔán-kʔo'dal / ʔkañ-ʔko-dal*
Loc.: C, SW, W, & S TX; common in Travis Co.; *D. gypsophila* also in TX (Culberson Co. only).
Form: small tree, up to 30 ft. tall; perennial.
Flowers: Feb-Apr (blue, purple; strong fragrance comparable to artificial grape).
Family: Fabaceae (Legume Family).
WARNING – DEADLY POISONOUS. One bean's contents may be

sufficient to cause death. The beans contain the alkaloid sophorine / cytisine (Castetter and Opler 1936, Schultes 1969, Vestal and Schultes 1939). This is one of the most poisonous plants in the area. Its beans are very hard, bright red to yellow, sometimes with peach or black mottling. There are no edible wild beans in the area that are red.

Toxicoscordion nuttallii (A.Gray) Rydb.
Nuttall's deathcamas

= Toxicoscordion texense, Zigadenus nuttallii, Z. texensis, Zygadenus texensis
Gosiute: *ta'-bĭ-si-go-ûp*
Loc.: C & NE TX; common in Travis Co; only *Toxicoscordion* sp. in TX.
Form: grass-like herb; perennial.
Family: Melanthiaceae (Bunchflower Family).

WARNING – DEADLY POISONOUS. The whole plant is highly toxic, especially the bulb, which can cause fatal poisoning, along with burning pain in the mouth, nausea, profuse vomiting, foaming at the mouth, dizziness, and mania, with death expected if the patient does not vomit freely.

Look-alikes – This plant looks similar to edible lily species, especially *Camassia scilloides*. The flowers of *Toxicoscordion nuttallii* are white, whereas those of *C. scilloides* are a lavender or purple color.

Left: *Dermatophyllum secundiflorum* foliage, bean pods, harvested beans. Range map: GBIF.
Right: *Toxicoscordion nuttallii* flowering plant in late March. Range map: GBIF.

Toxicodendron radicans (L.) Kuntze
Eastern poison ivy

= Rhus radicans, R. toxicodendron var. radicans
Cherokee: *higinà^{li}í* – "my friend" / *u^{d}lɔdá* / *u'ladá* / *u'ladá*, Diné: *k'ic'ictjjic*, Ojibwe: *anîmîki'bûg* – "cloud," Forest Potawatomi: *makaki'bag* – "toad weed," Prairie Potawatomi: *tatapa'kwe* – "climbs trees"
Loc.: all TX except NW, far S, & far N; very common in Travis Co.
Form: low shrub, shrub, vine.
Family: Anacardiaceae (Sumac Family).

WARNING – DEADLY POISONOUS. Consumption of any part of the plant may cause an allergic reaction resulting in asphyxiation and death. Urushiol, the oil in *Toxicodendron* spp., which causes dermatitis, induces the immune system to attack skin cells to which the urushiol binds rather than directly causing a reaction.

Urushiol is most concentrated on the surface of the green leaves, but one can get dermatitis from contact with the stems or other woody parts. As an oil, it can be cleaned off with water and soap within 30 minutes of contact. After this, the urushiol will have permanently bound to the skin, and cannot be cleaned off. The treatment of poison ivy dermatitis is mostly to relieve the itching sensation. As it is an autoimmune response, the trajectory of healing is mostly fixed.

Look-alikes – *Toxicodendron* fruits cannot be confused with *Rhus* (sumac) fruits, as the former are smooth, small, whitish to greenish, and borne hanging, whereas those of sumac are fuzzy, larger, yellow to red, and borne more erect. Poison sumac (*Toxicodendron vernix*) is only in far east Texas. Boxelder (*Acer negundo*) has leaves very similar to eastern poison ivy but they are arranged opposite, whereas poison ivy leaves are alternate.

Notes – Herbal treatments from plants in Texas that were historically used by Indigenous North Americans to treat poison ivy dermatitis include a decoction or poultice of *Artemisia ludoviciana* leaves (Northern Cheyenne, Mexican Kickapoo) or an infusion of *Capsella bursa-pastoris* or *Lepidium virginicum* plants (Menominee).

Round Valley Natives used the sap or juice of *Toxicodendron diversilobum* to remove warts: the wart was cut out, sap applied to the root, and re-applied several times over one or two days, after which it was gone. The sap was used in a similar manner to remove ringworm. The Kiowa rubbed the leaves or sap of *T. radicans* on boils, skin, eruptions, and other types of running or non-healing sores. Dermatitis followed, and as it healed, so did the affliction. The Hocąk and Prairie Potawatomi employed similar treatments. These uses likely relied on up-regulation of the immune system at the site of urushiol binding.

Toxicodendron radicans. Base photo: ten-foot-tall shrub fruiting in mid-July (note it can be a vine, scandent, or erect. Overlay, clockwise from top left: range map (GBIF); typical leaf; variant with no lobes and large leaves; variant with exceptionally numerous lobes and large leaves; bark and adventitious roots; green fruits (maturing white to light brown) in mid-June; flowers in mid-April.

INDEX OF SPECIES

Scientific name	Main	Photo	Scientific name	Main	Photo
Acer negundo	62	64	*Chenopodium berlandieri*	189	191
Agave spp.	236		*Chenopodium giganteum*	189	191
Agave americana	243	244	*Cicuta maculata*	273	273
Agave lechuguilla	243	245	*Colocasia esculenta*	274	275
Allium spp.	216		*Commelina erecta*	219	219
Allium canadense	218	220	*Condalia hookeri*	161	161
Allium drummondii	218	221	*Conium maculatum*	273	273
Amaranthus spp.	181		*Cylindropuntia leptocaulis*	52	54
Amaranthus albus	183		*Dasylirion* spp.	248	
Amaranthus blitoides	183	186	*Dasylirion texanum*	250	251
Amaranthus cruentus	184	186	*Datura wrightii*	275	275
Amaranthus palmeri	184	185	*Dermatophyllum secundflorum*	275	276
Amaranthus retroflexus	184		*Diospyros texana*	84	88
Ampelopsis cordata	174	180	*Diospyros virginiana*	86	87
Arbutus xalapensis	89	90	*Ehretia anacua*	79	80
Artemisia ludoviciana	198	198	*Elymus canadensis*	269	
Asclepias spp.	192		*Elymus virginicus*		271
Asclepias asperula	195	196	*Eriochloa sericea*	269	
Asclepias oenotheroides	195	196	*Festuca octoflora*	269	
Asclepias texana	195	196	*Forestiera pubescens*	141	142
Asclepias tuberosa	195		*Fraxinus pennsylvanica*	143	144
Asclepias verticillata	195		*Hordeum murinum*	269	
Asclepias viridiflora	195		*Ilex vomitoria*	73	76
Asclepias viridis	195	196	*Juglans* spp.	131	
Avena fatua	268	271	*Juglans major*	133	
Berberis trifoliolata	268	270	*Juglans microcarpa*	133	134
Bromus diandrus	269	271	*Juglans nigra*	133	134
Camassia scilloides	246	247	*Juniperus* spp.	47	
Capsicum annuum glabriusculum	210	212	*Juniperus ashei*	49	50
Carya illinoinensis	126	130	*Juniperus virginiana*	49	50
Celtis laevigata	81	83	*Liatris punctata*	197	199
Cercis canadensis	91	92	*Lindera benzoin*	135	135
Chenopodium spp.	186		*Lycoperdon* spp.	41	
Chenopodium album	189	190	*Lycoperdon perlatum*	42	43

Lycoperdon pyriforme	42		*Prunus angustifolia*	153	155
Malvaviscus arboreus	154	154	*Prunus mexicana*	153	156
Monarda spp.	200		*Prunus serotina*	153	157
Monarda citriodora	201	202	*Quercus* spp.	110	
Monarda fistulosa	201		*Quercus buckleyi*	122	
Monarda punctata	202		*Quercus fusiformis*	122	124
Morus spp.	136		*Quercus macrocarpa*	122	125
Morus alba	137	139	*Quercus marilandica*	122	
Morus microphylla	137	139	*Quercus muehlenbergii*	122	125
Morus rubra	138	140	*Quercus sinuata*	123	
Muhlenbergia spp.	269		*Quercus stellata*	123	
Muhlenbergia lindheimeri		271	*Quercus virginiana*	123	
Nekemias arborea	174	180	*Rhus* spp.	68	
Neltuma glandulosa	93	107	*Rhus aromatica*	70	71
Nelumbo lutea	228	230	*Rhus lanceolata*	70	72
Nolina spp.	252		*Rhus trilobata*	70	71
Nolina lindheimeriana	253	253	*Rhus virens*	70	71
Nolina texana	253	254	*Rubus* spp.	163	
Opuntia spp.	55		*Rubus trivialis*	165	166
Opuntia engelmannii	60	61	*Sabal* spp.	261	
Opuntia ficus-indica	60		*Sabal mexicana*	262	264
Opuntia humifusa	60		*Sabal minor*	262	263
Oxalis spp.	203		*Sambucus* spp.	65	
Oxalis dillenii	205	206	*Sambucus canadensis*	66	67
Oxalis drummondii	205	206	*Sarcomphalus obtusifolius*	147	148
Panicum spp.	270		*Schoenoplectus* spp.	224	
Panicum capillare		272	*Schoenoplectus californicus*	226	227
Parkinsonia aculeata	108	109	*Schoenoplectus pungens*	226	
Phytolacca americana	145	146	*Schoenoplectus tabernaemontani*	226	
Plantago spp.	207		*Setaria* spp.	270	
Plantago rhodosperma	208	209	*Sideroxylon lanuginosum*	160	162
Plantago wrightiana	208	209	*Smilax* spp.	167	
Pleurotus ostreatus	44	45	*Smilax bona-nox*	169	170
Poa spp.	270		*Smilax rotundifolia*	169	171
Poaceae	265		*Smilax tamnoides*	169	172
Populus deltoides	158	159	*Sorghum halapense*	270	272
Prunus spp.	149		*Sporobolus* spp.	170	

Tinantia anomala	219	219	*Urtica chamaedryoides*	214	215	
Toxicodendron radicans	277	278	*Urtica gracilis*	214	215	
Toxicoscordion nuttallii	276	276	*Vitis* spp.	173		
Tradescantia spp.	222		*Vitis aestivalis*	177	179	
Tradescantia gigantea	222	223	*Vitis berlandieri*	177	179	
Tradescantia occidentalis	222	223	*Vitis cinerea*	177	179	
Tradescantia ohiensis	223		*Vitis monticola*	177	179	
Tradescantia virginiana	223		*Vitis mustangensis*	177	179	
Trametes veriscolor	46	46	*Yucca* spp.	255		
Typha spp.	231		*Yucca arkansana*	258	258	
Typha domingensis	234	235	*Yucca rupicola*	258	259	
Typha latifolia	234		*Yucca treculiana*	258	260	
Urtica spp.	213					

Common name	Main	Photo	Common name	Main	Photo
African-spinach (red amaranth)	184	186	Bulrush, softstem	226	
Agarita	268	270	Camas, large	246	247
Agave	236		Carelessweed (Palmer amaranth)	184	185
Agave, American (century plant)	243	244	Cattail	231	
Anacua	79	80	Cattail, broadleaf	234	
Ash, green	143	144	Cattail, southern	234	235
Barley, mouse	269		Cherry, black	153	157
Beargrass	252		Chiltepín (chili pequín)	210	212
Beargrass, Texas	253	254	Common threesquare	226	
Beebalm	200		Cottonwood, eastern	158	159
Beebalm, lemon	201	202	Cupgrass, Texas	269	
Beebalm, spotted	202		Dayflower, false	219	219
Beebalm, wildbergamot	201		Dayflower, whitemouth	219	219
Blackberry	163		Deathcamas, Nuttall's	276	276
Blazing star, dotted	197	199	Devil's shoestring	253	253
Bluegrass	270		Dewberry	165	166
Bluejacket	223		Dropseed	170	
Boxelder	62	64	Elderberry	65	
Brazilian bluewood	161	161	Elderberry, American black	66	67
Bristlegrass	270		Fescue, sixweeks	269	
Brome, ripgut	269	271	Garlic, meadow	218	220
Bulrush	224		Goosefoot	186	
Bulrush, California	226	227	Goosefoot, pitted	189	191

281

Grape	173	
Grape, Berlandier's	177	179
Grape, graybark	177	179
Grape, mustang	177	178
Grape, summer	177	179
Grape, sweet mountain	177	179
Grass	265	
Green antelopehorn	195	196
Greenbrier	169	
Greenbrier, bristly	169	172
Greenbrier, roundleaf (bullbrier)	169	171
Greenbrier, saw (catbrier)	169	170
Gum bumelia	160	162
Hemlock, poisonous	273	273
Hemlock, spotted water	273	273
Indian fig	60	
Johnsongrass	270	272
Juniper (cedar)	47	
Juniper, Ashe (mountain cedar)	49	50
Juniper, red (eastern redcedar)	49	50
Lamb's-quarter	189	190
Lechuguilla	243	245
Lotebush	147	148
Lotus, American	228	230
Madrone, Texas	89	90
Mesquite, honey	93	107
Milkweed	192	
Milkweed, butterfly	195	
Milkweed, eastern whorled	195	
Milkweed, green comet	195	
Milkweed, spider	195	196
Milkweed, Texas	195	196
Milkweed, zizotes	195	196
Muhly	269	
Muhly, Lindheimer's		271
Mulberry	136	
Mulberry, red	138	140
Mulberry, Texas	137	139
Mulberry, white	137	139
Nettle	213	
Nettle, heartleaf	214	215
Nettle, stinging	214	215
Oak	110	
Oak, bastard	123	
Oak, blackjack	122	
Oak, bur	122	125
Oak, chinquapin	122	125
Oak, live	123	
Oak, post	123	
Oak, Texas live	122	124
Oak, Texas red	122	
Oat, wild	268	270
Onion (garlic)	216	
Onion, Drummond's	218	221
Oyster mushroom	44	45
Palmetto	261	
Palmetto, dwarf	262	269
Palmetto, Mexican	262	264
Paloverde (retama)	108	109
Panicgrass	270	
Pecan	126	130
Peppervine	174	180
Peppervine, heartleaf	174	180
Persimmon, common	86	87
Persimmon, Texas	84	88
Pigweed (amaranth)	181	
Pigweed, prostate	183	
Pigweed, redroot	184	
Pigweed, spreading	183	186
Plantain	207	
Plantain, redseed	208	209
Plantain, Wright's	208	209
Plum (cherry)	149	
Plum, Chickasaw	153	155

Plum, Mexican	153	156		Sumac, flameleaf	70	72
Poison ivy, eastern	277	278		Sumac, fragrant	70	71
Pokeweed	145	146		Sumac, skunkbush	70	71
Prickly pear (nopal)	55			Taro (elephant ears)	274	275
Prickly pear, Engelmann's	60	61		Tasajillo (Christmas cholla)	52	54
Prickly pear, low (devil's tongue)	60			Texas mountain laurel	275	276
Puffball	41			Tree spinach	189	191
Puffball, common	42	43		Turkey tail	46	46
Puffball, pear-shaped	42			Walnut	131	
Redbud	91	92		Walnut, Arizona black	133	
Sacred thorn-apple (datura)	275	275		Walnut, eastern black	133	134
Sagebrush, white	198	198		Walnut, little	133	134
Sotol	248			Wax mallow (Turk's cap)	154	154
Sotol, Texas	250	251		Wildrye, Canada	269	
Spanish dagger	258	260		Wildrye, Virginia		271
Spicebush, northern	135	135		Witchgrass		272
Spiderwort	222			Woodsorrel	203	203
Spiderwort, giant	222	223		Woodsorrel, Drummond's	205	206
Spiderwort, prairie	222	223		Woodsorrel, slender yellow	205	206
Spiderwort, Virginia	223			Yaupon	73	76
Stretchberry (elbowbush)	141	142		Yucca	255	
Sugarberry (hackberry)	81	83		Yucca, Arkansas	258	258
Sumac	68			Yucca, twisted-leaf	258	259
Sumac, evergreen	70	71				

REFERENCES

INDEX OF PEOPLES BY ETHNOGRAPHIC REFERENCES

PEOPLE	REFERENCES	PEOPLE	REFERENCES
Acansa	Foster 2008	Asinai	Foster 1998
American Indians	Havard 1895	Asinai	Hatcher 1927a-d
American Indians	Mason 1894	Asinai	Kress et al. 1931
American Indians	Palmer 1871	Asinai	Ohlendorf et al. 1980
Apache	Bandelier 1890	Assiniboine	Blair 1911
Apache	Basso and Opler 1971	Atákapa	Dyer 1917
Apache	Battey 1875	Atákapa	Mayhall 1939
Apache	Bell and Castetter 1937	Blackfeet	Hellson and Gadd 1974
Apache	Bourke 1895	Blackfeet	McClintock 1909
Apache	Castetter and Opler 1936	Caddo	Abel 1922
Apache	Forrestal 1935	Caddo	Berlandier 1969
Apache	Mails 1974	Caddo	Catlin 1852
Apache	Mason 1894	Caddo	Fritz 1993
Apache	Palmer 1871	Caddo	Hatcher 1927a-d
Apache	San Carlos Apache College 2020	Caddo	Kress et al. 1931
Apache, Chiricahua	Castetter and Opler 1936	Caddo	La Vere 2006
Apache, Jicarilla	Robbins et al. 1916	Caddo	Ohlendorf et al. 1980
Apache, Lipan	Banta 1911	Caddo	Swanton 1996
Apache, Lipan	Ohlendorf et al. 1980	Cáhita	Beals 1943
Apache, Mescalero	Castetter and Opler 1936	Cahuilla	Barrows 1967
Apache, Mescalero	Mails 1974	Cahuilla	Bean and Saubel 1972
Apache, Western	Basso and Opler 1971	California Natives	Hammett and Lawlor 2004
Apache, Western	Castetter and Opler 1936	California Natives	Heizer and Elsasser 1980
Apache, Western	Mails 1974	California Natives	Jacknis 2004
Aranama	Celiz and Hoffmann 1935	California Natives	Lightfoot and Parrish 2009
Arapaho	Cowell 2005	California Natives	Mead 1972
Arapaho	Schultes 1969	California Natives	Powers 1877
Arizona Natives	Palmer 1871	California Natives, Baja	Clavijero 1852
Arkansas Natives	Palmer 1871	California Natives, Northern	Chesnut 1902
Arkansas Natives in 1540	Varner and Varner 1988	California Natives, Southern	Baegert 1979
Asinai	Berlandier 1969	California Natives, Southern	Beals 1943
Asinai	Cunningham 2006	California Natives, Southern	Willson and Caughey 1995

284

Calpella	Chesnut 1902	Gosiute	Chamberlin 1911
Cape Flattery and Pitt River Natives	Palmer 1871	Great Basin Mojave Desert	Fowler 1995
Carrizo	Forrestal 1931	Great Lakes Natives	Blair 1911
Carolina Natives in 1700	Hudson 1990	Haudenosaunee	Parker 1910
Cenis	Weddle 1987	Haudenosaunee	Waugh 1916
Chemehuevi	Fowler 1995	Haudenosaunee, Cayuga	Waugh 1916
Cherokee	Banks 1953	Haudenosaunee, Mohawk	Waugh 1916
Cherokee	Hamel and Chiltoskey 1975	Haudenosaunee, Onondaga	Waugh 1916
Cherokee	Perdue 1993	Haudenosaunee, Seneca	Waugh 1916
Cheyenne, Northern	Bernier 2004	Haudenosaunee, Tonawanda	Waugh 1916
Cheyenne, Northern	Hart 1981	Hidatsa	Kurz 1970
Choctaw	Bushnell 1909	Hocąk	Kidscher and Hurlburt 1998
Choctaw	Perdue 1993	Hocąk	Radin 1970
Choctaw	Swanton 2001	Hoopa	Havard 1895
Choctaw	Vestal and Schultes 1939	Hopi	Bell and Castetter 1937
Coahuiltecans	Ohlendorf et al. 1980	Hopi	Fewkes 1896
Cocopah	Bell and Castetter 1937	Hopi	Hough 1897
Cocopah	Garcés 1900	Hopi	Hough 1898
Cocopah	Gifford 1933	Hopi	Palmer 1871
Cocopah	Kelly 1977	Hopi	Whiting 1939
Comanche	Battey 1875	Houma	Speck 1941
Comanche	Berlandier 1969	Hualapai	Palmer 1871
Comanche	Carlson and Jones 1939	Huron	Waugh 1916
Comanche	Kavanagh 2008	Illinois	Waugh 1916
Comanche	La Vere 2006	Illinois Natives	Palmer 1871
Concow	Chesnut 1902	Iowa	Schultes 1969
Cree	Leighton 1985	Isleta Pueblo	Jones 1931
Creek	Bushnell 1909	Jemez Pueblo	Cook 1930
Crow	Blankenship 1905	Kamia	Gifford 1931
Dakota	Gilmore 1977	Kansas Natives in 1540	Winship 1904
Dakota, Santee	Gilmore 1977	Karankawa	Berlandier 1969
Dakota, Teton	Gilmore 1977	Karankawa	Gatschet 1891
Diné	Elmore 1944	Karankawa	Ohlendorf et al. 1980
Diné	Navajo Traditional Teachings 2019	Karuk	Schenck and Gifford 1952
First Nations	Palmer 1871	Kawaiisu	Zigmond 1981
Flathead	Hart 1979	Kawaiisu	Zigmond 1981
Florida Natives in 1540	Varner and Varner 1988	Kichai	La Vere 2006
		Kickapoo, Mexican	Latorre and Latorre 1977

Kiliwa	Beals 1943	Natchitoches	Hatcher 1927a-d
Kiowa	Battey 1875	Natchitoches	Kress et al. 1931
Kiowa	La Vere 2006	Natchitoches	Swanton 1996
Kiowa	Schultes 1937	Nazones	Hatcher 1927a-d
Kiowa	Vestal and Schultes 1939	Nazones	Kress et al. 1931
Klamath	Coville 1897	Nebome	Bandelier 1890
Kumeyaay	Bolton 1930	New Mexico Natives	Palmer 1871
Lakota	Munson 1981	New Spain, Indigenous	Hernando 1970
La Salle expedition to Texas	Foster 1998	Nez Perce	Havard 1895
Little Lake	Chesnut 1902	Nomlaki	Palmer 1871
Maidu, Northern	Dixon 1905	Oglala	Gilmore 1977
Maricopa	Bell and Castetter 1937	Ojibwe	Densmore 1928
Mayo	Bandelier 1890	Ojibwe	Smith 1932
Menominee	Blair 1911	Oklahoma Natives	Palmer 1871
Menominee	Smith 1923	Omaha	Gilmore 1977
Mescaleros in Mexico	Ohlendorf et al. 1980	Opata	Bourke 1895
Meskwaki	Smith 1928	Osage	Munson 1981
Mexican folk	Havard 1895	Pacific Northwest Natives	French 1965
Mexican folk, Tamaulipas	Ohlendorf et al. 1980	Pacific Northwest Natives	Palmer 1871
Mexican folk	Palmer 1871	Paiute, Northern	Fowler 1990
Mexican folk	Zigmond 1981	Paiute, Southern	Fowler 1995
Mexican folk	Ximenez 1888	Pawnee	Gilmore 1977
Mexican folk, Northern	Bourke 1895	Pawnee	Palmer 1871
Mexican, Indigenous	Ebeling 1986	Pima	Bell and Castetter 1937
Mexican, Indigenous	Vestal and Schultes 1939	Pima	Bolton 1930
Mexican, Indigenous, in 1500s	Castañeda 1904	Pima	Bourke 1895
Mexican, Indigenous, in 1519	MacNutt 1908	Pima	Jones 1931
Mississippi Valley, Upper	Blair 1911	Pima	Rea 1997
Miwok, Sierra	Heizer and Elsasser 1980	Pima, Gila River	Rea 1997
Moapa	Fowler 1995	Pima	Whittemore 1893
Mohave	Barrows 1967	Plains tribes	Dodge 1959
Mohave	Bell and Castetter 1937	Plains tribes	Ebeling 1986
Mohave	Hrdlička 1908	Pomo	Chesnut 1902
Mojave Desert Natives	Fowler 1995	Pomo, Kashaya	Goodrich et al. 1980
Montana tribes	Blankenship 1905	Ponca	Gilmore 1977
Myaamia	Everest et al. 2019	Potawatomi	Smith 1933
Nabedache	Abel 1922	Potawatomi, Forest	Smith 1933
Nabedache	Hatcher 1927a-d	Potawatomi, Prairie	Smith 1933
Nabedache	Kress et al. 1931	Pueblo	Palmer 1871
Nabedache	Swanton 1996	Pueblo	Jones 1931

Puebloans in 1540	Winship 1904	Texians	Berlandier 1969
Quapaw	Foster 1998	Texians	Ohlendorf et al. 1980
Round Valley Natives	Chesnut 1902	Tewa	Robbins et al. 1916
Salteaux	Blair 1911	Timbasha	Coville 1892
Santa Fe Natives in 1645	Perez de Ribas 1645	Timbasha	Fowler 1995
Sauk	Blair 1911	Tohono O'odham	Bourke 1895
Seri	Felger and Moser 1985	Tohono O'odham	Castetter and Underhill 1935
Shasta	Dixon 1907	Tohono O'odham	Hrdlička 1908
Shoshone	Irwin 1980	Tohono O'odham	Palmer 1871
Sioux	Palmer 1871	Tolowa	Baker 1981
Sonoran Natives, prehistoric	Felger and Moser 1985	Tonkawa	Berlandier 1969
South Carolina Natives in 1567	Hudson 1990	Tonkawa	Celiz and Hoffmann 1935
South Carolina Natives in 1540	Varner and Varner 1988	Tonkawa	Mayhall 1939
Southeast Natives	Battle 1922	Tonkawa	Ohlendorf et al. 1980
Southeast Natives	Bushnell 1909	Ute	Palmer 1871
Southeast Natives	Gatschet 1891	Ute, Northern	Smith 1974
Southeast Natives	Havard 1895	Wailaki	Chesnut 1902
Southwest Natives	Bell and Castetter 1941	Walapai	Bell and Castetter 1937
Southwest Natives	Bourke 1895	Western Natives	Ebeling 1986
Southwest Natives	Palmer 1871	Wichita	La Vere 2006
Taos	Bell and Castetter 1941	Yavapai	Gifford 1932
Tawokani	Berlandier 1969	Yokia	Chesnut 1902
Tawokani	Ohlendorf et al. 1980	Yokuts	Gayton 1948
Tejon	Zigmond 1981	Yuki	Chesnut 1902
Tepehuán	Bandelier 1890	Yuma	Barrows 1967
Texas Natives	Berlandier 1969	Yuma	Bell and Castetter 1937
Texas Natives	Ohlendorf et al. 1980	Yuma	Forde 1931
Texas Natives	Havard 1895	Yuma	Garcés 1900
Texas Natives, Central	Ohlendorf et al. 1980	Yuma	Heintzelman 1853
Texas Natives in 1530	Augenbraum and Cabeza de Vaca 2013	Yurok	Baker 1981
Texas Natives in 1709	Tous and Foik 1930	Zuñi	Bell and Castetter 1941
Texas Natives in the Archaic	Bell and Castetter 1941	Zuñi	Stevenson 1909
Texas Natives in the Archaic	Dering 1999	Zuñi	Winship 1904
Texas Natives in the Archaic (5,000 ya)	Williams-Dean 1978		

Alternate names for peoples referenced in text:

PEOPLE	ALT. NAME	PEOPLE	ALT. NAME
American Indians	Native Americans	Isleta	Tiwa
American Indians	Other Natives	Jemez	Towa
Apache	Ndé	Kamia	Kumeyaay
Asinai	Tejas	Karuk	Karok
California Natives, Northern	Calpella, Concow, Little Lake, Nomlaki, Pomo, Wailaki, Yokia, Yuki	Kawaiisu	Nüwa
California Natives, Southern	Natives of Lower California in 1852	Kichai	Keechi
Cherokee	Tsalagi	Kumeyaay	Kumeyaay Indians of San Sebastian, CA in 1775
Comanche	Nʉmʉnʉʉ	Mayo	Yoreme
Diné	Navajo	New Spain, Indigenous	Indians of New Spain in 1790
Haudenosaunee	Iroquois	Ojibwe	Chippewa
Hidatsa	Hiraacá	Round Valley Natives	Calpella, Concow, Little Lake, Nomlaki, Pomo, Wailaki, Yokia, Yuki
Hocąk	Ho-Chunk	Sioux	Oceti Sakowin (Dakota, Lakota)
Hoopa	Hupa	Texians	Texas folk under Mexican rule
Hopi	Moqui	Timbasha	Panamint
Illinois	Illiniwek	Tohono O'odham	Papago
Iowa	Báxoje	Zuñi	A:shiwi

Note that this is not a complete list of all alternative names historically or contemporarily used for the peoples referenced. Many groups have been recorded under different spellings, exonyms, or descriptive labels over time, and some names used in early sources may no longer be in common use or may be considered inaccurate or inappropriate today. This table provides a representative sampling to aid in cross-referencing sources and understanding historical accounts, but it does not capture the full diversity of names that have appeared in ethnographic, linguistic, or historical records.

ETHNOGRAPHIC AND HISTORICAL REFERENCES

Abel, Annie Helois, ed. 1922. *A report from Natchitoches in 1807 by Dr. John Sibley.* New York: Museum of the American Indian.

Anderson, Gary C. 1999. *The Indian Southwest, 1580-1830: ethnogenesis and reinvention.* Norman: University of Oklahoma Press.

Anderson, Kat M. 2005. *Tending the wild: Native American knowledge and the management of California's natural resources.* Berkeley: University of California Press.

Augenbraum, Harold, ed., and Álvar Núñez Cabeza de Vaca. 2013. *Narrative of the Narváez expedition.* Chicago: Lakeside Press, R. R. Donnelley & Sons Company.

Baegert, Johann Jakob, S.J. 1979. *Observations in Lower California.* Berkeley: University of California Press.

Baker, Marc Andre. 1981. *The ethnobotany of the Yurok, Tolowa, and Karok Indians of northwest California.* M.A. thesis, Humboldt State University, Arcata, CA.

Bandelier, A. F. 1890. *Final report of investigations among the Indians of the southwestern United States, carried on mainly in the years from 1880 to 1885.* American Series III. Cambridge: Papers of the Archaeological Institute of America, Cambridge University Press.

Banks, William H. 1953. *Ethnobotany of the Cherokee Indians.* M.S. thesis, University of Tennessee, Knoxville, TN.

Banta, S. E. 1911. *Buckelew the Indian Captive: or the life story of F. M. Buckelew while a captive among the Lipan Indians in the western wilds of frontier Texas as related by himself.* Mason, TX: The Mason Herald.

Barrett, S. A., and E. W. Gifford. 1933. "Miwok material culture." *Bulletin of Milwaukee Public Museum* 2(4):118–388.

Barrows, David Prescott. 1967. *The ethno-botany of the Coahuilla Indians of Southern California.* Banning, CA: Malki Museum Press.

Basso, Keith H., and Morris E. Opler, eds. 1971. *Apachean culture history and ethnology.* Tucson: University of Arizona Press.

Battey, Thomas C. 1875. *Life and adventures of a Quaker among the Indians.* Boston: Lee and Shepard, Publishers.

Battle, Herbert B. 1922. "The domestic use of oil among the southern Aborigines." *American Anthropologist* 24(2):171–182.

Beals, Ralph L. 1943. *The aboriginal culture of the Cáhita Indians.* Ibero-Americana 19. Berkeley: University of California Press.

Bean, John Lowell, and Katherine Siva Saubel. 1972. *Temalpakh (from the earth): Cahuilla Indian knowledge and usage of plants.* Banning, CA: Malki Museum Press.

Bell, William H., and Edward F. Castetter. 1937. *The utilization of mesquite and screwbean by Aborigenes in the American Southwest.* Ethnobiological Studies in the American Southwest, Biological Series 5(2). Albuquerque: University of New Mexico Press.

Bell, Willis H., and Edward F. Castetter. 1941. *The utilization of yucca, sotol, and beargrass by the aborigines in the American Southwest.* Biological Series 5(5). Albuquerque: University of New Mexico Press.

Berlandier, Jean Louis. 1969. *The Indians of Texas in 1830.* Edited by John C. Ewers, translated by Patricia Reading Leclerq. Washington, DC: Smithsonian Institution Press.

Bernier, Gabriel Ruben. 2004. *Ethnobotany of the Northern Cheyenne: medicinal plants.* M.A. thesis, University of Montana, Missoula, MT.

Blair, Emma Helen, transl. and ed. 1911. *The Indian Tribes of the Upper Mississippi Valey and region of the Great Lakes: as described by Nicolas Perrot, French commandant in the Northwest; Bacqueville de la Potherie, French royal commissioner to Canada; Morrel Marston, American army officer; and Thomas Forsyth, United States agent at Fort Armstrong.* Vol. 1. Cleveland: The Arthur H. Clark Company.

Blankinship, J. W. 1905. "Native economic plants of Montana." In *An ethnobiology source book: the uses of plants and animals by American Indians*, edited by David Hurst Thomas, 1986. New York: Garland Publishing, Inc.

Bolton, Herbert Eugene, transl. and ed. 1930. *Anza's California expeditions 4: Font's complete diary of the second Anza expedition.* Berkeley: University of California Press.

Bourke, John Gregory. 1895. *Folk-foods of the Rio Grande Valley of northern Mexico.* Boston: American Folk-Lore Society, Houghton, Mifflin & Co.

Bushnell, David I., Jr. 1909. *The Choctaw of Bayou Lacomb, St. Tammany Parish, Louisiana.* Bureau of American Ethnology Bulletin 48. Washington, DC: Smithsonian Institution.

Carlson, Gustav G., and Volney H. Jones. 1939. "Some notes on uses of plants by the Comanche Indians." *Papers of the Michigan Academy of Science, Arts, and Letters* 25:517–542.

Castañeda, 1904. *The journey of Coronado, 1540–1542: from the City of Mexico to the Grand Canon of the Colorado and the buffalo plains of Texas, Kansas, and Nebraska as told by himself and his followers.* Translated and edited by George Parker Winship. New York: A. S. Barnes & Company.

Castetter, Edward F., and M. E. Opler. 1936. *The ethnobiology of the Chiricahua and Mescalero Apache: the use of plants for foods, beverages, and narcotics.* Ethnobiological Studies in the American Southwest, Biological Series 4(5). Albuquerque: University of New Mexico Press.

Castetter, Edward F., and Ruth M. Underhill. 1935. *The ethnobiology of the Papago Indians.* Ethnobiological Studies in the American Southwest 2, The University of New Mexico Bulletin 275, Biological Series 4(3). Albuquerque: University of New Mexico Press.

Catlin, George. 1852. *Caddo Indians Gathering Wild Grapes.* Tulsa: Gilcrease Museum. https://collections.gilcrease.org/object/012148 (accessed April 22, 2019).

Celiz, Francisco, and Fritz Leo Hoffmann, transl. 1935. *Diary of the Alarcon expedition into Texas, 1718–1719.* Los Angeles: The Quivira Society.

Chamberlin, Ralph V. 1911. "The ethno-botany of the Gosiute Indians of Utah." *Memoirs of the American Anthropological Association* 2(5):330–384.

Chesnut, V. K. 1902. *Plants used by the Indians of Mendocino County, California.* Washington, DC: Government Printing Office.

Clavijero, Francisco Javier. 1852. *Historia de la Antigua ó Baja California*. Translated by Nicolas Garcia De San Vicente. Edited by Juan R. Navarro. Mexico: Imprenta De Juan R. Navarro.

Cook, Sarah Louise. 1930. *The ethnobotany of Jemez Indians*. M.A. thesis, University of New Mexico.

Cordain, Loren, J. B. Miller, S. B. Eaton, N. Mann, S. H. A. Holt, and J. D. Speth. 2000. "Plant-animal subsistence ratios and macronutrient energy estimations in worldwide hunter-gatherer diets." *The American Journal of Clinical Nutrition* 71(3):682–692.

Coville, Frederick Vernon. 1892. *The Panamint Indians of California*. Washington, DC: The American Anthropologist V. Judd & Detweiler, Printers.

Coville, Frederick Vernon. 1897. "Notes on the plants used by the Klamath Indians of Oregon." *Contributions from the U.S. National Herbarium* 5(2):86–109. Washington, DC: Government Printing Office.

Crown, Patricia L., Thomas E. Emerson, Jiyan Gu, W. Jeffrey Hurst, Timothy R. Peuketat, and Timothy Ward. 2012. "Ritual Black Drink consumption at Cahokia." *PNAS* 109(35):13944–13949.

Cunningham, Debbie S. 2006. "The Domingo Ramón diary of the 1716 expedition into the province of the Tejas Indians: an annotated translation." *The Southwestern Historical Quarterly* 110(1):38–67.

De Bry, Theodor. 1591. *Brevis narratio eorum quae in Florida Americae provincia Gallis acciderunt anno 1564: Secunda pars Americae*. Plate XX, "Ceremony of the Cassina Drink." Frankfurt: Johann Wechel.

Densmore, Frances. 1928. *Uses of plants by the Chippewa Indians*. Bureau of American Ethnology 44:276–397.

Dering, Phil. 1999. "Earth-oven plant processing in Archaic Period economies: an example from a semi-arid savannah in south-central North America." *American Antiquity* 64(4):659–674.

Dixon, Roland B. 1905. "The Huntington California expedition: the Northern Maidu." *Bulletin of the American Museum of Natural History* 17(part III):119–346.

Dixon, Roland B. 1907. *The Shasta*. American Museum of Natural History Bulletin 17(5).

Dodge, Richard Irving. 1959. *The Plains of the great West and their inhabitants; being a description of the Plains, game, Indians, &c. of the great North American desert*. New York: Archer House.

Dyer, J. O. 1917. *The Lake Charles Atakapas (cannibals) period of 1817 to 1820*. Galveston, TX.

Ebeling, Walter. 1986. *Handbook of Indian Foods and Fibers of Arid America*. Berkeley: University of California Press.

Elmore, Francis H. 1943. *Ethnobotany of the Navajo*. Monograph Series 1(7). Albuquerque: University of New Mexico Press.

Everest, Michael A., Michael P. Gonella, Holly G. Bowler, and Joshua R. Waschak. 2019. "How toxic is milkweed when harvested and cooked according to Myaamia tradition?" *Ethnobiology Letters* 10(1):50–56.

Felger, Richard Stephen, and Mary Beck Moser. 1985. *People of the desert and sea: ethnobotany of the Seri Indians*. Tucson: The University of Arizona Press.

Fewkes, J. Walter. 1896. "A contribution to ethnobotany." *The American Anthropologist* 9(1):14–21.

Forde, Daryll C. 1931. *Ethnography of the Yuma Indians.* University of California Publications in American Archaeology and Ethnology 28(4):83–278. Berkeley: University of California Press.

Forrestal, Peter P. 1931. "The Solis diary of 1767." *Preliminary Studies of the Texas Catholic Historical Society* 1(6).

Forrestal, Peter P. 1935. "Peña's diary of the Aguayo Expedition." *Preliminary Studies of the Texas Catholic Historical Society* 2(7).

Foster, William C. 1995. *Spanish expeditions into Texas, 1689–1768.* Austin: University of Texas Press.

Foster, William C., ed. 1998. *The La Salle expedition to Texas: the journal of Henri Joutel, 1684–1687.* Translated by Johanna S. Warren. Austin: Texas State Historical Association.

Foster, William C. 2008. *Historic Native Peoples of Texas.* Austin: University of Texas Press.

Foster, William C. 2012. *Climate and culture change in North America AD 900–1600.* Austin: University of Texas Press.

Francis, James Eric Sr. 2008. "Burnt Harvest: Penobscot people and fire." *Maine History* 44(1):4–18.

Fowler, Catherine S. 1990. *Tule technology: Northern Paiute uses of Marsh resources in western Nevada.* Washington, DC: Smithsonian Institution Press.

Fowler, Catherine S. 1995. "Some notes on ethnographic subsistence systems in Mojavean environments in the Great Basin." *Journal of Ethnobiology* 15(1):99–117.

Franklin, Ethel Mary, and Annie P. Harris. 1937. "Memoirs of Mrs. Annie P. Harris." *The Southwestern Historical Quarterly* 40(3):231–246.

French, David H. 1965. "Ethnobotany of the Pacific Northwest Indians." *Economic Botany* 19(4):378–382.

Fritz, Gayle K. 1993. "Archaeobotanical analysis." In *Caddoan saltmakers in the Ouachita valley: the Hardman site*, edited by Ann M. Early, Arkansas Archaeological Survey Research Series 44. Fayetteville, AR.

Garcés, Francisco. 1900. *On the trail of a Spanish pioneer: the diary and itinerary of Francisco Garcés in his travels through Sonora, Arizona, and California 1775–1776.* New York: Francis P. Harper.

Gatschet, Albert S. 1891. *The Karankawa Indians: the coast people of Texas.* Archaeological and Ethnological Papers of the Peabody Museum 1(2). Salem, MA: The Salem Press Publishing and Printing Co.

Gayton, A. H. 1948. "Yokuts and Western Mono ethnography I: Tulare Lake, Southern Valley, and Central Foothill Yokuts." *Anthropological Records* 10(2):143–302. Berkeley: University of California Press.

Gifford, Edward Winslow. 1931. *The Kamia of Imperial Valley.* Smithsonian Institution Bureau of American Ethnology Bulletin 97. Washington, DC: Government Printing Office.

Gifford, Edward Winslow. 1932. *The Southeastern Yavapai.* University of California Publications in American Archaeology and Ethnology 29(3):177–252. Berkeley: University of California Press.

Gifford, Edward Winslow. 1933. *The Cocopa.* University of California Publications in American Archaeology and Ethnology 31(5):257–334. Berkeley: University of California Press.

Gilmore, Melvin Randolph. 1977. *Uses of plants by the Indians of the Missouri River region.* Lincoln: University of Nebraska Press.

Goodrich, J., C. Lawson, and V. P. Lawson. 1980. *Kashaya Pomo plants.* Los Angeles: American Indian Studies Center, University of California.

Hamel, Paul B., and Mary U. Chiltoskey. 1975. *Cherokee Plants and Their Uses: A 400 Year History.* Sylva, NC: Herald Publishing Co.

Hammett, Julia E., and Elizabeth J. Lawlor. 2004. "Paleoethnobotany in California." In *People and plants in ancient western North America*, edited by Paul E. Minnis. Washington, DC: Smithsonian Books.

Harp, Cyrus. 2014a. "Edible insect use by California Indians." Poster presented at the Society of Ethnobiology and Society of Economic Botany joint annual conference, Cherokee, NC, May 14.

Harp, Cyrus. 2014b. "Insect foods of the California Indians." Guest lecture, *Insects in Human Society* (undergraduate course, prof. Kip Will), UC Berkeley, CA, Nov. 11.

Hart, Jeffrey A. 1979. "The ethnobotany of the Flathead Indians of western Montana." *Botanical Museum Leaflets* 27(10):261–307. Cambridge, MA: Harvard University.

Hart, Jeffrey A. 1981. "The ethnobotany of the Northern Cheyenne Indians of Montana." *Journal of Ethnopharmacology* 4:1–55.

Hatcher, Mattie Austin. 1927a. "Description of the Tejas or Asinai Indians, 1671–1722, I." *The Southwestern Historical Quarterly* 30(3):206–218. Austin: Texas State Historical Association.

Hatcher, Mattie Austin. 1927b. "Description of the Tejas or Asinai Indians, 1671–1722, II." *The Southwestern Historical Quarterly* 30(4):283–304. Austin: Texas State Historical Association.

Hatcher, Mattie Austin. 1927c. "Description of the Tejas or Asinai Indians, 1671–1722, III." *The Southwestern Historical Quarterly* 31(1):50–62. Austin: Texas State Historical Association.

Hatcher, Mattie Austin. 1927d. "Description of the Tejas or Asinai Indians, 1671–1722, IV." *The Southwestern Historical Quarterly* 31(2):150–180. Austin: Texas State Historical Association.

Havard, V. 1895. "Food plants of the North American Indians." *Bulletin of the Torrey Botanical Club* 22.

Heintzelman, Samuel P. 1853. "Official report of Samuel P. Heintzelman." *Journal of California and Great Basin Anthropology* 28(1):89–102.

Heizer, Robert F., and Albert B. Elsasser. 1980. *The natural world of the California Indians.* Berkeley: University of California Press.

Hellson, John C., and Morgan Gadd. 1974. *Ethnobotany of the Blackfoot Indians.* Mercury Series, Canadian Ethnology Service Paper 19. Ottawa: National Museums of Canada.

Hernando, Francisco. 1790. *Medici atque historici* 1: de historia plantarum novae Hispaniae. Matriti: Ex Typographia Ibarrae Heredum. Modified from the edition of 1651.

Hough, Walter. 1897. "The Hopi in relation to their plant environment." *American Anthropologist* 10(2):33–47.

Hough, Walter. 1898. "Environmental interrelations in Arizona." *American Anthropologist* 11(5):133–155.

Hrdlička, Aleš. 1908. *Physiological and medical observations: among the Indians of southwestern United States and northern Mexico.* Bureau of American Ethnology 34(8):1–427. Washington, DC: Government Printing Office.

Hudson, Charles. 1990. *The Juan Pardo expeditions: explorations of the Carolinas and Tennessee, 1566–1568.* Tuscaloosa: University of Alabama Press.

Irwin, Charles N., ed. 1980. *The Shoshoni Indians of Inyo County, California: the Kerr manuscript.* Ballena Press Publications in Archaeology, Ethnology, and History 15. Edited by Philip J. Wilke and Albert B. Elsasser.

Jacknis, I. 2004. "Notes toward a culinary anthropology of Native California." In *Food in California Indian culture*, edited by I. Jacknis. Berkeley: Phoebe Apperson Hearst Museum of Anthropology.

Jones, Volney H. 1931. *The ethnobotany of the Isleta Indians.* M.A. Thesis, University of New Mexico.

Karttunen, Frances. 1992. *An analytical dictionary of Nahuatl.* Norman: University of Oklahoma Press.

Kavanagh, Thomas W., ed. 2008. *Comanche ethnography: field notes of E. Adamson Hoebel, Waldo R. Wedel, Gustav G. Carlson, and Robert H. Lowie.* Lincoln: University of Nebraska Press.

Kelly, William H. 1977. *Cocopa ethnography.* Anthropological Papers of the University of Arizona 29. Tucson: University of Arizona Press.

Kindscher, Kelly, and Dana P. Hurlburt. 1998. "Huron Smith's ethnobotany of the Hocąk (Winnebago)." *Economic Botany* 52:352–372.

Kress, Margaret Kenney, Fray Gaspar José de Solís, and Mattie Austin Hatcher. 1931. "Diary of a visit of inspection of the Texas Missions made by Fray Gaspar José de Solís in the year 1767–68." *Southwestern Historical Quarterly* 35(1):28–76.

Kurz, Rudolph Friederich. 1970. *Journal of Rudolph Friederich Kurz: An account of his experiences among fur traders and American Indians on the Mississippi and the Upper Missouri Rivers during the years 1846 to 1852.* Translated by Myrtis Jarrell. Edited by J.N.B. Hewitt. Lincoln: University of Nebraska Press.

Latorre, Dolores L., and Felipe A. Latorre. 1977. "Plants used by the Mexican Kickapoo Indians." *Economic Botany* 31(3):340–357.

La Vere, David. 2006. *Life among the Texas Indians: the WPA narratives.* College Station: Texas A&M University Press.

Leighton, Anna L. 1985. "Wild plant use by the Woods Cree (Nihīthawak) of east-central Saskatchewan." Canadian Ethnology Service Paper No. 101. Ottawa: National Museum of Man, National Museums of Canada.

Lightfoot, Kent G., and Otis Parrish. 2009. *California Indians and their environment.* Berkeley: University of California Press.

MacNutt, Francis A., trans. and ed. 1908. *Letters of Cortes: the five letters of relation from Fernando Cortes to the Emperor Charles V.* Vol. 1. New York: G. P. Putnam's Sons.

Mails, Thomas E. 1974. *People called Apache.* Englewood Cliffs, NJ: Prentice-Hall, Inc.

Malo, Davida. 1903. *Hawaiian antiquities (Moolelo Hawaii).* Honolulu: Honolulu Gazette Co., Ltd.

Mason, Otis T. 1894. "North American bows, arrows, and quivers." *Smithsonian Report for 1893*:631–679. Washington, DC: Government Printing Office.

Mayhall, Mildred Pickle. 1939. *The Indians of Texas: the Atákapa, the Karankawa, the Tonkawa.* PhD diss., University of Texas, Austin.
McClintock, Walter. 1909. "Materia medica of the Blackfeet." *Zeitschrift für Ethnologie* 51:273–279.
Mead, George R. 1972. *The ethnobotany of the California Indians: a compendium of plants, their users, and their uses.* Greeley, CO: Museum of Anthropology, University of Northern Colorado.
Munson, Patrick J. 1981. "Contributions to Osage and Lakota ethnobotany." *Plains Anthropologist* 26(93):229–240.
Newcomb, W. W. 1993. *The Indians of Texas.* Austin: University of Texas Press.
Ohlendorf, Sheila M., Josette M. Bigelow, and Mary M. Standifer, eds. and trans. 1980. *Journey to Mexico during the years 1826 to 1834 by Jean Louis Berlandier.* Vol. 2. Austin: Texas State Historical Association.
Palmer, Edward. 1871. *Food products of the North American Indians.* United States Commissioner of Agriculture. Annual Report for 1870.
Parker, Arthur C. 1910. "Iroquois uses of maize and other food plants." *Museum Bulletin* 144. Albany: New York State Museum.
Perdue, Theda. 1993. *Nations remembered: an oral history of the Cherokees, Chickasaws, Choctaws, Creeks, and Seminoles in Oklahoma, 1865–1907.* Norman: University of Oklahoma Press.
Perez de Ribas, Andres. 1645. *Historia de los triumphos de nuestra Santa Fee entre gentes las mas barbaras y fieras del nuevo orbe: conseguidos por los soldados de la milicia de la compañia de Jesus en las missiones de la provincia de Nueva España.* Madrid: Por Aloso de Paredes.
Perttula, Timothy K. 2012. *The prehistory of Texas.* College Station: Texas A&M University Press.
Powers, Stephen. 1877. *Tribes of California.* Washington, DC: Government Printing Office.
Radin, Paul. 1970. *The Winnebago tribe.* Lincoln: University of Nebraska Press.
Rea, Amadeo M. 1997. *At the desert's green edge: an ethnobotany of the Gila River Pima.* Tucson: University of Arizona Press.
Robbins, Wilfred William, John Peabody Harrington, and Barbara Freire-Marreco. 1916. *Ethnobotany of the Tewa Indians.* Bureau of American Ethnology Bulletin 55. Washington, DC: Government Printing Office.
San Carlos Apache College. 2020, December 4. "Traditional Apache Foods (Nest'an) [Video]." YouTube. https://youtu.be/jArEa_mXtYQ.
Schenck, Sara M., and E. W. Gifford. 1952. "Karok ethnobotany." *Anthropological Records* 13(6):377–392. Berkeley: University of California Press.
Schultes, Richard Evans. 1937. "Peyote and plants used in the Peyote Ceremony." *Harvard University Botanical Museum Leaflets* 4(8):129–152.
Schultes, Richard Evans. 1969. "Hallucinogens of plant origin." *Science,* New Series 163(3864):245–254.
Smith, Ann M. 1974. *Ethnography of the Northern Utes.* Papers in Anthropology No. 17. Santa Fe: Museum of New Mexico Press.
Smith, Huron H. 1923. "Ethnobotany of the Menomini Indians." *Bulletin of the Public Museum of the City of Milwaukee* 4(1):1–174.
Smith, Huron H. 1928. "Ethnobotany of the Meskwaki Indians." *Bulletin of the Public Museum of the City of Milwaukee* 4(2):175–326.
Smith, Huron H. 1932. "Ethnobotany of the Ojibwe Indians." *Bulletin of the Public Museum of the City of Milwaukee* 4(3):327–525.

Smith, Huron H. 1933. "Ethnobotany of the Forest Potawatomi Indians." *Bulletin of the Public Museum of the City of Milwaukee* 7(1):1–230.

Speck, Frank G. 1941. "A list of plant curatives obtained from the Houma Indians of Louisiana." *Primitive Man* 14(4):49–75.

Stevenson, Matilda Coxe. 1909. *Ethnobotany of the Zuñi Indians.* Thirtieth annual report of the Bureau of American Ethnology.

Sturtevant, William C., ed. 1979. *Handbook of North American Indians: Volume 9: Southwest.* Washington, DC: Smithsonian Institution.

Swanton, John Reed. 1996. *Source material on the history and ethnology of the Caddo Indians.* Norman: University of Oklahoma Press.

Swanton, John Reed. 2001. *Source material for the social and ceremonial life of the Choctaw Indians.* Tuscaloosa: University of Alabama Press.

Tous, Gabriel, and Paul J. Foik, eds. 1930. "The Espinosa-Olivares-Aguirre expedition of 1709." *Preliminary Studies of the Texas Catholic Historical Society* 1(3).

Varner, John Grier, and Jeannette Johnson Varner, trans. and eds. 1988. *The Florida of the Inca: a history of the Adelantado, Hernando de Soto, Governor and Captain General of the kingdom of Florida, and of other heroic Spanish and Indian cavaliers, written by The Inca, Carcilaso de la Vega, an officer of His Majesty, and a native of the great city of Cuzco, capital of the realms and provinces of Peru.* Austin: University of Texas Press.

Vestal, Paul A., and Richard Evans Schultes. 1939. *Economic botany of the Kiowa Indians: as it relates to the history of the tribe.* Cambridge, MA: Botanical Museum.

Waugh, F. W. 1916. *Iroquis [sic] foods and food preparation.* Geological Survey Memoir 86(12). Ottawa, Canada: Canada Department of Mines, Government Printing Bureau.

Weddle, Robert S., ed. 1987. *La Salle, the Mississippi, and the Gulf.* College Station: Texas A&M University Press.

Whiting, Alfred F. 1939. *Ethnobotany of the Hopi.* Flagstaff: Museum of Northern Arizona.

Whittemore, Isaac T. 1893. *Among the Pimas: the mission to the Pima and Maricopa Indians.* Albany, NY: The Ladies' Union Mission School Association.

Williams-Dean, Glenna Joyce. 1978. *Ethnobotany and cultural ecology of prehistoric man in southwest Texas.* PhD diss., Texas A&M University.

Winship, George Parker, ed. 1904. *The Coronado Expedition, 1540-1542.* New York: A. S. Barnes & Co.

Ximenez, F. 1888. *Cuatro libros de la naturaleza y virtudes de las plantas y animales, de uso medicinal en la Nueva España – Mexico.* 1st ed. Mexico: Oficina tip. de la Secretaria de fomento, 1615.

Zigmond, Maurice L. 1981. *Kawaiisu ethnobotany.* Salt Lake City: University of Utah Press.

BOTANICAL AND OTHER REFERENCES

1) Abbasi, A. M., M. H. Shah, X. Guo, and N. Khan. 2016. "Comparison of nutritional value, antioxidant potential, and risk assessment of the mulberry (*Morus*) fruits." *International Journal of Fruit Science* 16(2): 113–34.
2) Acuña, Laura I. 2006. *The economic contribution of root foods and other geophytes in prehistoric Texas.* M.A. thesis, Texas State University, San Marcos, TX.
3) Aiton, William, Franz Andreas Bauer, James Sowerby, Georg Dionysius Ehret, and George Nicol. 1789. *Hortus Kewensis, or, a catalogue of the plants cultivated in the Royal Botanic Garden at Kew.* Vol. 1. London: Printed for George Nicol, Bookseller to His Majesty.
4) Amadi, B. A., U. C. Njoku, E. N. Agumuo, P. U. Amadi, O. E. Ezendiokwere, and K. T. Nwauche. 2017. "Assessment of vitamins, protein quality and mineral bioavailability of matured stems of *Opuntia dillenii* grown in Nigeria." *Bioengineering and Bioscience* 5(3): 47–54.
5) Amaral, J. S., S. Casal, J. A. Pereira, R. M. Seabra, and B. P. P. Oliveira. 2003. "Determination of sterol and fatty acid compositions, oxidative stability, and nutritional value of six walnut (*Juglans regia* L.) cultivars grown in Portugal." *Journal of Agricultural and Food Chemistry* 51: 7698–7702.
6) Amaral, J. S., M. R. Alves, R. M. Seabra, and B. P. P. Oliveira. 2005. "Vitamin E composition of walnuts (*Juglans regia* L.): A 3-year comparative study of different cultivars." *Journal of Agricultural and Food Chemistry* 53(13): 5467–72.
7) Antora, S. A., K.-V. Ho, C.-H. Lin, A. L. Thomas, S. T. Lovell, and K. Krishnaswamy. 2022. "Quantification of vitamins, minerals, and amino acids in black walnut (*Juglans nigra*)." *Frontiers in Nutrition* 9: 936189.
8) Appenteng, et al. 2021. "Cyanogenic glycoside analysis in American elderberry." *Molecules* 26(5): 1384.
9) Arndt, S. K., S. C. Clifford, and M. Popp. 2001. "*Ziziphus*—A multipurpose fruit tree for arid regions." In *Sustainable land use in deserts*, 388–99. Berlin, Heidelberg: Springer Berlin Heidelberg.
10) Atanasov, A. G., S. M. Sabharanjak, G. Zengin, A. Mollica, A. Szostak, M. Simirgiotis, Ł. Huminiecki, O. K. Horbańczuk, S. M. Nabavi, and A. Mocan. 2017. "Pecan nuts: A review of reported bioactivities and health effects." *Trends in Food Science & Technology* 71: 246–57.
11) Bhusal, K. K., S. K. Magar, R. Thapa, A. Lamsal, S. Bhandari, R. Maharjan, S. Shrestha, and J. Shrestha. 2022. "Nutritional and pharmacological importance of stinging nettle (*Urtica dioica* L.): A review." *Heliyon* 8(6): e09717.
12) Boyd, M. L., and P. J. Cotty. 2001. "*Aspergillus flavus* and aflatoxin contamination of leguminous trees of the Sonoran Desert in Arizona." *Phytopathology* 91(9): 913–19.
13) Buczyński, K., M. Kapłan, and Z. Jarosz. 2024. "Review of the report on the nutritional and health-promoting values of species of the *Rubus* L. genus." *Agriculture* 14(8): 1324.
14) Chaouali, N., I. Gana, A. Dorra, F. Khelifi, A. Nouioui, W. Masri, I. Belwaer, H. Ghorbel, and A. Hedhili. 2013. "Potential toxic levels of cyanide in almonds (*Prunus amygdalus*), apricot kernels (*Prunus armeniaca*), and almond syrup." *International Scholarly Research Notices* 2013: 610648.
15) Cheatham, Scooter, Marshall C. Johnston, and Lynn Marshall. 1995. *The useful wild plants of Texas, the Southeastern and Southwestern United States, the*

Southern Plains and Northern Mexico. Vol. 3. Austin, TX: Useful Wild Plants, Inc.
16) Chen, S. Y., and T. M. T. Rozaina. 2020. "Effect of Cooking Methods on the Nutritional Composition and Antioxidant Properties of Lotus (*Nelumbo nucifera*) Rhizome." *Food Research* 4(4):1207–1216.
17) Clemente-Villalba, Jesús, Francisco Burló, Francisca Hernández, and Ángel A. Carbonell-Barrachina. 2024. "Potential Interest of *Oxalis pes-caprae* L., a Wild Edible Plant, for the Food and Pharmaceutical Industries." *Foods* 13(6):858.
18) Crown, Patricia L., Thomas E. Emerson, Jiyan Gu, W. Jeffrey Hurst, Timothy R. Pauketat, and Timothy Ward. 2012. "Ritual Black Drink Consumption at Cahokia." *Proceedings of the National Academy of Sciences* 109(35):13944–13949.
19) Cruz-Requena, M., H. De La Garza-Toledo, C. N. Aguilar-González, A. Aguilera-Carbó, H. Reyes-Valdés, and M. Rutiaga. 2013. "Chemical and Molecular Properties of Sotol Plants (*Dasylirion cedrosanum*) of Different Sex and Its Fermentation Products." *International Journal of Basic and Applied Chemical Sciences* 3:41–49.
20) Demir, F., H. Doğan, M. Özcan, and H. Haciseferoğullari. 2002. "Nutritional and Physical Properties of Hackberry (*Celtis australis* L.)." *Journal of Food Engineering* 54(3):241–247.
21) Di Minin, Enrico, Nigel Leader-Williams, and Corey J. A. Bradshaw. 2016. "Banning Trophy Hunting Will Exacerbate Biodiversity Loss." *Trends in Ecology & Evolution* 31(2):99–102.
22) Dorman, H. J. D., and S. G. Deans. 2004. "Chemical Composition, Antimicrobial and in Vitro Antioxidant Properties of *Monarda citriodora* var. *citriodora*, *Myristica fragrans*, *Origanum vulgare* ssp. *hirtum*, *Pelargonium* sp. and *Thymus zygis* Oils." *Journal of Essential Oil Research* 16(2):145–150.
23) Dulger, Basaran. 2005. "Antimicrobial Activity of Ten Lycoperdaceae." *Fitoterapia* 76:352–354.
24) Edwards, Adam L., and Bradley C. Bennett. 2005. "Diversity of Methylxanthine Content in *Ilex cassine* L. and *Ilex vomitoria* Ait.: Assessing Sources of the North American Stimulant Cassina." *Economic Botany* 59(3):275–285.
25) Everest, Michael A., Michael P. Gonella, Holly G. Bowler, and Joshua R. Waschak. 2019. "How Toxic Is Milkweed When Harvested and Cooked According to Myaamia Tradition?" *Ethnobiology Letters* 10(1):50–56.
26) Everitt, J. H., M. A. Alaniz. 1981. "Nutrient Content of Cactus and Woody Plant Fruits Eaten by Birds and Mammals in South Texas." *The Southwestern Naturalist* 26(3):301.
27) Foster, Steven, and James A. Duke. 2000. *A Field Guide to Medicinal Plants and Herbs of Eastern and Central North America.* New York: Houghton Mifflin.
28) Granett, Jeffrey, M. Andrew Walker, Laszlo Kocsis, and Amir D. Omer. 2001. "Biology and Management of Grape Phylloxera." *Annual Review of Entomology* 46:387–412.
29) Greene, Robert A. 1932. "Composition of the Beans of *Parkinsonia aculeata*." *Botanical Gazette* 94(2):411–415.
30) Guerrero, José Luis Guil, and María Esperanza Torija Isasa. 1997. "Nutritional Composition of Leaves of *Chenopodium* Species (*C. album* L., *C. murale* L. and *C. opulifolum* Shraeder)." *International Journal of Food Sciences and Nutrition* 48:321–327.

31) Harden, M. L., and R. Zolfaghari. 1988. "Nutritive Composition of Green and Ripe Pods of Honey Mesquite (*Prosopis glandulosa*, Fabaceae)." *Economic Botany* 42(4):522–532.
32) Iglesias, Antía, Ángeles Cancela, Xana Álvarez, and Ángel Sánchez. 2023. "Anthocyanins and total phenolic compounds from pigment extractions of non-native species from the Umia River Basin: *Eucalyptus globulus*, *Tradescantia fluminensis*, and *Arundo donax*." *Applied Sciences* 13(10):5909.
33) Janzen, D. H., and P. S. Martin. 1982. "Neotropical anachronisms: the fruit the gomphotheres ate." *Science* 215:19–27.
34) Ji, Yunheng, Jacob B. Landis, Jin Yang, Shuying Wang, Nian Zhou, Yan Luo, and Haiyang Liu. 2023. "Phylogeny and evolution of Asparagaceae subfamily Nolinoideae: new insights from plastid phylogenomics." *Annals of Botany* 131(2):301–312.
35) Jones, Stanley D., Joseph K. Wipff, and Paul M. Montgomery. 1997. *Vascular plants of Texas: a comprehensive checklist including synonymy, bibliography, and index.* Austin, TX: University of Texas Press.
36) King, Steven R., and Stanley N. Gershoff. 1987. "Nutritional evaluation of three underexploited Andean tubers: *Oxalis tuberosa* (Oxalidaceae), *Ullucus tuberosus* (Basellaceae), and *Tropaeolum tuberosum* (Tropaeolaceae)." *Economic Botany* 41(4):503–511.
37) Loveridge, Andrew J., Jonathan C. Reynolds, and E. J. Milner-Gulland. 2009. "Does sport hunting benefit conservation?" In *Key Topics in Conservation Biology*, edited by David Macdonald and Katrina Service. Malden, MA: Blackwell Publishing.
38) Luna-Vázquez, F. J., C. Ibarra-Alvarado, A. Rojas-Molina, J. I. Rojas-Molina, E. M. Yahia, D. M. Rivera-Pastrana, A. Rojas-Molina, and Á. M. Zavala-Sánchez. 2013. "Nutraceutical value of black cherry *Prunus serotina* Ehrh. fruits: antioxidant and antihypertensive properties." *Molecules* 18(12):14597–14612.
39) MacFarlane, A. M. 1924. *Gaelic plant names: study of their uses and lore.* Inverness: The Northern Chronicle.
40) Metzler, Susan, and Van Metzler. 1992. *Texas mushrooms: a field guide.* Austin, TX: University of Texas Press.
41) Miller, Allison J., David A. Young, and Jun Wen. 2001. "Phylogeny and biogeography of *Rhus* (Anacardiaceae) based on ITS sequence data." *International Journal of Plant Sciences* 162(6):1401–1407.
42) Monroy-García, I. N., L. L. González-Galván, C. Leos-Rivas, M. Z. Treviño-Garza, E. Sánchez-García, and E. Viveros-Valdez. 2025. "In vitro gastrointestinal bioaccessibility of the phenolic fraction from *Agave inaequidens* flower." *Foods* 14(13):2375.
43) Morris, Elizabeth Ann, W. Max Witkind, Ralph L. Dix, and Judith Jacobson. 1981. "Nutritional content of selected aboriginal foods in northeastern Colorado: buffalo (*Bison bison*) and wild onions (*Allium* spp.)." *Journal of Ethnobiology* 1(2):213–220.
44) Murage, Joyce W. 2020. *Analysis of lycopene, vitamin A and beta carotene in red cactus (Opuntia ficus-indica) fruit in Nyeri County, Kenya.* PhD diss., Kenyatta University, Kenya.
45) Olvera-Fonseca, Silvia. 2004. "Evaluation of the bromatological potential of seeds and fruits of *Sabal mexicana* Mart. (Arecaceae)." *Economic Botany* 58(4):536–543.
46) Orozco-Sifuentes, M. M., M. H. Reyes-Valdes, and J. E. García. 2019. "Nutritive potential of sotol (*Dasylirion cedrosanum*) seeds." *Revista Fitotecnia Mexicana* 42(4):385–392.

47) Ota, A., A. M. Višnjevec, R. Vidrih, Ž. Prgomet, M. Nečemer, J. Hribar, N. G. Cimerman, S. S. Možina, M. Bučar-Miklavčič, and N. P. Ulrih. 2017. "Nutritional, antioxidative, and antimicrobial analysis of the Mediterranean hackberry (*Celtis australis* L.)." *Food Science & Nutrition* 5(1):160–170.
48) Pareek, Sunil. 2013. Nutritional composition of jujube fruit. *Emirates Journal of Food & Agriculture* 25(6):463-470.
49) Paulson, Nels. 2012. The place of hunters in global conservation advocacy. *Conservation & Society* 10(1):53-62.
50) Pío-León, J. F., S. P. Díaz-Camacho, M. G. López, J. Montes-Ávila, G. López-Ángulo, and F. Delgado-Vargas. 2012. Características fisicoquímicas, nutricias y antioxidantes del fruto de *Ehretia tinifolia*. *Revista Mexicana de Biodiversidad* 83(1):273–280.
51) Písaříková, B., J. Peterka, M. Trčková, J. Moudrý, Z. Zralý, and I. Herzig. 2006. Chemical composition of the above-ground biomass of *Amaranthus cruentus* and *A. hypochondriacus*. *Acta Vet. Brno* 75:133-138.
52) Poonia, Amrita, and Ashutosh Upadhayay. 2015. *Chenopodium album* Linn: review of nutritive value and biological properties. *Journal of Food Science and Technology* 52(7):3977-3985.
53) Prakash, Dhan, Pashupati Nath, and M. Pal. 1993. Composition, variation of nutritional contents in leaves, seed protein, fat and fatty acid profile of *Chenopodium* species. *J. Sci. Food Agric.* 62:203-205.
54) Repo-Carrasco, R., C. Espinoza, and S.-E. Jacobsen. 2003. Nutritional value and use of the Andean crops quinoa (*Chenopodium quinoa*) and kañiwa (*Chenopodium pallidicaule*). *Food Reviews International* 19(1&2):179-189.
55) Senica, M., I. Kump, M. Stampar, and F. Mikulic-Petkovsek. 2017. The higher the better? Differences in phenolics and cyanogenic glycosides in *Sambucus nigra* leaves, flowers and berries from different altitudes. *J. Sci. Food Agric.* 97(8):2623-2632.
56) Serfass, Thomas L., Robert P. Brooks, and Jeremy T. Bruskotter. 2018. North American model of wildlife conservation: empowerment and exclusivity hinder advances in wildlife conservation. *Canadian Wildlife Biology & Management* 7(2):101-118.
57) Silva, S., E. M. Costa, A. Borges, A. P. Carvalho, M. J. Monteiro, and M. M. E. Pintado. 2016. Nutritional characterization of acorn flour (a traditional component of the Mediterranean gastronomical folklore). *Journal of Food Measurement and Characterization* 10(3):584–588.
58) Šircelj, Helena, Maja Mikulič-Petkovšek, and Franc Batič. 2010. Antioxidants in spring leaves of *Oxalis acetosella* L. *Food Chemistry* 123(2):351–357.
59) Smith, B. N., G. K. Hoops, L. S. Land, and R. L. Folk. 1971. 13C/12C ratios of aragonite from fruit of *Celtis laevigata* Willd. *Die Naturwissenschaften* 58(7):365–366.
60) Soria-Melgarejo, Gonzalo, Juan Raya-Peréz, Juan Ramírez-Pimentel, Glenda Gutiérrez-Benicio, Isaac Andrade-González, and Cesar Aguirre-Mancilla. 2024. Physicochemical, nutritional properties, and antioxidant potential of 'limilla' fruit (*Rhus aromatica* var. *schmidelioides* (Schltdl.) Engl.). *Heliyon* 10:e34990.
61) Stronza, Amanda L., Carter A. Hunt, and Lee A. Fitzgerald. 2019. Ecotourism for conservation? *Annual Review of Environment and Resources* 44:229-253.
62) Sun, J., Q. Li, J. Li, J. Liu, and F. Xu. 2022. Nutritional composition and antioxidant properties of the fruit of *Berberis heteropoda* Schrenk. *PLOS ONE* 17(4):e0262622.

63) Tanwar, Beenu, Rajni Mogdil, and Ankit Goyal. 2021. Nutritional and phytochemical composition of pecan nut [*Carya illinoinensis* (Wangenh.) K. Koch] and its hypocholesterolemic effect in an animal model. *British Food Journal* 123(4):1433-1448.
64) Thapa, R., M. Edwards, and M. W. Blair. 2021. Relationship of cultivated grain amaranth species and wild relative accessions. *Genes* 12(12):1849.
65) Thoms, Alston V. 2008. Ancient Savannah Roots of the Carbohydrate Revolution in South-Central North America. *Plains Anthropologist* 53(205):121–136.
66) Tucker, A. O., M. J. Maciarella, P. W. Burbage, and G. Sturtz. 1994. Spicebush [*Lindera benzoin* (L.) Blume var. *benzoin*, Lauraceae]: a tea, spice, and medicine. *Economic Botany* 48(3):333–336.
67) Tull, Delena. 1987. *Edible and useful plants of Texas and the Southwest*. University of Texas Press, Austin, TX.
68) U.S. Department of Agriculture, Agricultural Research Service. 2019. Agave, cooked (Southwest) (FDC ID 35193). *FoodData Central*. Accessed August 30, 2025. https://fdc.nal.usda.gov/food-details/35193/nutrients.
69) U.S. Department of Agriculture, Agricultural Research Service. 2025. Grapes, red, seedless, raw (FDC ID 2346412). *FoodData Central*. Accessed August 27, 2025. https://fdc.nal.usda.gov/food-details/2346412/nutrients.
70) U.S. Department of Agriculture, Agricultural Research Service. 2025. Pecans (FDC ID 2346395). *FoodData Central*. Accessed August 25, 2025. https://fdc.nal.usda.gov/food-details/2346395/nutrients.
71) U.S. Department of Agriculture, Agricultural Research Service. 2025. Spices, pepper, red or cayenne (FDC ID 170932). *FoodData Central*. Accessed August 26, 2025. https://fdc.nal.usda.gov/food-details/170932/nutrients.
72) Vines, Robert A. 1960. *Trees, shrubs, and woody vines of the Southwest*. University of Texas Press, Austin, TX.
73) Vorderbruggen, Mark. 2022. *Foraging: explore nature's bounty and turn your foraged finds into flavorful feasts*. Dorling Kindersley Ltd., Indianapolis, IN.
74) Vu, L., and Q. K. Huynh. 1994. Isolation and characterization of a 27-kDa antifungal protein from the fruits of *Diospyros texana*. *Biochemical and Biophysical Research Communications* 202(2):666–672.
75) Wolfe, Wendy S., and Charles W. Weber. 1985. Use and nutrient composition of traditional Navajo foods. *Ecology of Food and Nutrition* 17(4):323-344.
76) Yi, Tingshuang, Allison J. Miller, and Jun Wen. 2004. Phylogenetic and biogeographic diversification of *Rhus* (Anacardiaceae) in the Northern Hemisphere. *Molecular Phylogenetics and Evolution* 33:861-879.
77) Zannou, Oscar, Kouame F. Oussou, Ifagbémi B. Chabi, Fadel Alamou, Nour M.H. Awad, Yann E. Miassi, Fifamè C.V. Loké, Adam Abdoulaye, Hojjat Pashazadeh, Ali Ali Redha, Yénoukounmè E. Kpoclou, Gamze Guclu, Ilkay Koca, Serkan Selli, and Salam A. Ibrahim. 2025. Phytochemical and nutritional properties of sumac (*Rhus coriaria*): a potential ingredient for developing functional foods. *Journal of Future Foods* 5(1):21-35.
78) Zecca, G., M. Labra, and F. Grassi. 2020. Untangling the evolution of American wild grapes: admixed species and how to find them. *Frontiers in Plant Science* 10:1814.
79) Zhang, Y., X. Lu, S. Zeng, X. Huang, Z. Guo, Y. Zheng, Y. Tian, and B. Zheng. 2015. Nutritional composition, physiological functions and processing of lotus (*Nelumbo nucifera* Gaertn.) seeds: a review. *Phytochemistry Reviews* 14:321–334.

DIGITAL DATABASES

Biota of North America Program (BONAP). *North American Plant Atlas (NAPA) / Taxonomic Data Center.* https://bonap.net.

Flora of North America Editorial Committee, eds. 1993–. *Flora of North America North of Mexico.* 25+ vols. New York and Oxford. https://floranorthamerica.org.

Global Biodiversity Information Facility. *GBIF.* https://www.gbif.org.
All range maps shown in this book are screen captures from GBIF. Several photos in this book, licensed in the Public Domain (CC0 1.0), are also from GBIF. These are *Trametes versicolor*, by "nodiflora" in Plano, TX, *Nelumbo lutea* by "d33j" in St. Louis, MO, and *Cicuta maculata* by Robert Hoard.

iNaturalist. https://www.inaturalist.org.

Integrated Taxonomic Information System. *ITIS.* https://www.itis.gov.

International Phonetic Association. *IPA.* https://www.internationalphoneticassociation.org.

International Plant Names Index. *IPNI.* https://www.ipni.org

Lady Bird Johnson Wildflower Center. *Native Plant Database.* Lady Bird Johnson Wildflower Center, The University of Texas at Austin. https://www.wildflower.org/plants/

Native Plant Society of Texas. *Travis County Flora Project.* https://www.npsot.org

Online Nahuatl Dictionary. *Wired Humanities Project.* https://nahuatl.wired-humanities.org.

Royal Botanic Gardens, Kew. *Plants of the World Online.* https://powo.science.kew.org.

Texas State Historical Association. *TSHA Online.* Austin, TX. https://www.tshaonline.org/.

Tropicos. *Missouri Botanical Garden.* https://www.tropicos.org.

USDA, NRCS. T*he PLANTS Database.* National Plant Data Team, Natural Resources Conservation Service, United States Department of Agriculture. https://plants.usda.gov

World Flora Online. *WFO.* https://www.worldfloraonline.org.

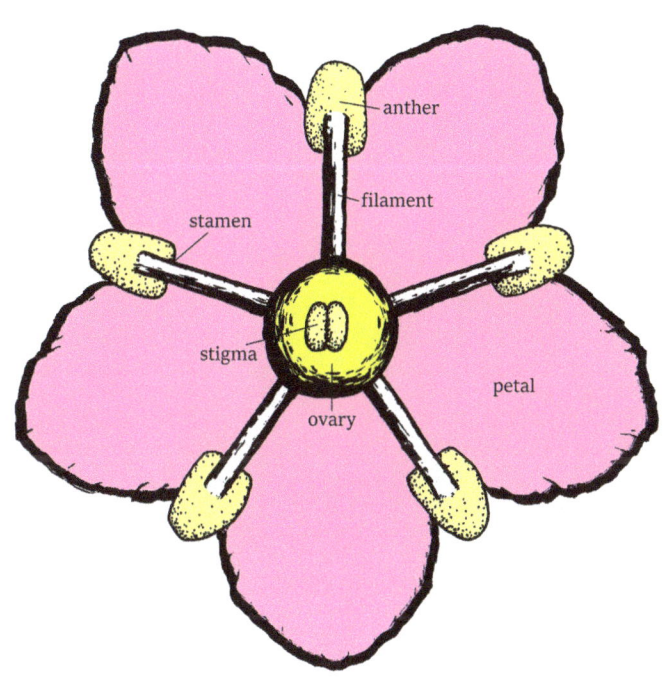

Thanks for reading!
Please appreciate nature, forage with respect, and preserve wild places.

Cyrus Harp
Founder of Paleo Foraging
PaleoForaging.com
@paleoforaging on YouTube and social media.

www.ingramcontent.com/pod-product-compliance
Lightning Source LLC
Chambersburg PA
CBHW052027030426
42337CB00027B/4902